# Management and Applications of Energy Storage Devices

*Edited by Kenneth E. Okedu*

Published in London, United Kingdom

IntechOpen

*Supporting open minds since 2005*

Management and Applications of Energy Storage Devices
http://dx.doi.org/10.5772/intechopen.94655
Edited by Kenneth E. Okedu

Contributors
Rushali R. Thakkar, Vijaya Kumar Kambila, Ruhul Amin, Marm Dixit, Nitin Muralitharan, Anand Parejiya, Rachid Essehli, Ilias Belharouak, Banavath Srinivasaraonaik, Shishir Sinha, Lok Pratap Singh, Ramasamy Velmurugan, Balasubramanian Subramanian, Tarun Kanti Bhattacharyya, Monojit Mondal, Arkaprava Datta, Efrén Fernández Palomeque, Diego Rojas Hiedra, Daniel Cordero, Martín Espinoza, Innocent E. Davidson, Rodney Reddy



First published in London, United Kingdom, 2022 by IntechOpen
IntechOpen is the global imprint of INTECHOPEN LIMITED, registered in England and Wales, registration number: 11086078, 5 Princes Gate Court, London, SW7 2QJ, United Kingdom
Printed in Croatia

British Library Cataloguing-in-Publication Data
A catalogue record for this book is available from the British Library

Additional hard and PDF copies can be obtained from orders@intechopen.com

Management and Applications of Energy Storage Devices
Edited by Kenneth E. Okedu
p. cm.
Print ISBN 978-1-83969-644-2
Online ISBN 978-1-83969-645-9
eBook (PDF) ISBN 978-1-83969-646-6

For EU product safety concerns:
IN TECH d.o.o., Prolaz Marije Krucifikse Kozulić 3, 51000 Rijeka, Croatia, info@intechopen.com or visit our website at intechopen.com.

# Meet the editor

Kenneth Eloghene Okedu was a research fellow in the Department of Electrical and Computer Engineering, Massachusetts Institute of Technology (MIT), Boston, USA, in 2013. He obtained his Ph.D. from the Department of Electrical and Electronic Engineering, Kitami Institute of Technology, Japan, in 2012. He received his BSc and MEng in Electrical and Electronic Engineering from the University of Port Harcourt, Nigeria, in 2003 and 2007, respectively, where he was retained as a faculty member from 2005 until the present day. He has also been a visiting faculty member at the Abu Dhabi National Oil Company (ADNOC) Petroleum Institute. He was also a visiting faculty member at the Caledonian College of Engineering, Oman (Glasgow Caledonian University, UK). He is presently a visiting professor in the Department of Electrical and Computer Engineering, National University of Science and Technology (NUST), Oman, and an adjunct professor in the Department of Electrical and Electronic Engineering, Nisantasi University, Turkey. He was recognized as a top 1% peer reviewer in Engineering by Publons in 2018 and 2019 and was the editor's pick in the *Journal of Renewable and Sustainable Energy* in 2018. Dr. Okedu has published several books and journals/transactions in the field of renewable energy. He is an editorial board member for journals including *Frontiers in Renewable Energy Research in Smart Grids*, *International Journal of Electrical Engineering*, *Mathematical Problems in Engineering*, and *Trends in Renewable Energy*. His research interests include power system stability, renewable energy systems, stabilization of wind farms, stability analysis of Doubly-Fed Induction Generators (DFIGs) and Permanent Magnet Synchronous Generator (PMSG) variable speed wind turbines, augmentation and integration of renewable energy into power systems, grid frequency dynamics, wind energy penetration, FACTS devices and power electronics, renewable energy storage systems, and hydrogen and fuel cells.

# Contents

# Preface

Renewable energy systems have been identified as potential solutions to traditional fossil fuels, which are associated with the emission of carbon dioxide that is not environmentally friendly. The 26th United Nations Climate Change Conference (COP26) in Glasgow, United Kingdom, brought parties together to accelerate actions towards the Paris Agreement and Kyoto Protocol.

With the proliferation of renewable energy and smart grid technologies, the need for energy storage cannot be overemphasized, due to the stochastic nature of renewable energy sources. In renewable energy applications, solar, wind, and hydropower energy sources are considered the most versatile for generating electricity. The high-efficiency power utilization of solar and wind energy, either standalone (autonomous) or grid-connected (non-autonomous), necessitates considerable advancements in energy storage technologies, especially in residential, industrial, solar farm, and wind farm applications, to overcome their intermittent nature.

There are emerging interdisciplinary studies on energy storage technologies for solar-photovoltaic, photothermal, electrical, and thermal energies, wind farm stability, and hydrogen production using battery energy storage systems, flywheels, or other alternative solutions to enhance the continuity and stability of solar and wind energy utilization and efficiency.

This book presents recent works on the concept, design, control, and applications of energy storage devices and techniques in the field of energy conversion and conservation, including electrical and thermal storage for the rapid development and utilization of energy systems. Since the developments in this field are rapid and accompanied by many challenges, this book proposes different techniques considering novel topologies of control and advanced strategies for the optimal performance of energy storage systems.

The book is for graduate and undergraduate students, academics, energy planners, industries, and government establishments in the field of energy storage devices and management.

This book is divided into two sections. The first section includes four chapters considering battery storage techniques. Chapter 1, "Electrical Equivalent Circuit Models of Lithium-ion Battery," helps in understanding the behavior of batteries in order to improve both performance and efficiency. A systematic comparison and analysis using simulation was done to help select an ideal model that best suits a specific application. Chapter 2, "Battery Energy Storage Systems and Rooftop Solar-Photovoltaics in Electric Power Distribution Networks," demonstrates the merits of a rooftop solar-photovoltaic (PV) system employing a battery energy storage system. It analyzes the application of only a battery energy storage system, only a rooftop PV system, a hybrid of the rooftop and battery energy system, and varying levels of the battery energy storage system. Chapter 3, "The Second Life of Hybrid Electric Vehicles Batteries Methodology of Implementation in Ecuador," studies the implementation of second-life NiHm batteries in hybrid vehicles in Ecuador to improve the quality of recycling of elements like battery packs and associated devices and generate a special treatment for their recovery. Chapter 4, "Current Status and Prospects of Solid-State Batteries as the Future of Energy Storage," discusses the specific challenges, design principles, and performance improvement strategies regarding the cathode, solid electrolyte, and anode employed in solid-state batteries. It emphasizes the various perspectives and outlooks on applications that could benefit from the implementation of solid-state battery systems.

The second section includes four chapters focusing on alternative energy storage techniques. Chapter 5, "Physicochemical Approaches for Thin Film Energy Storage Devices through PVD Techniques," discusses the growth of economical and eco-friendly thin, flexible supercap batteries and their potential in transporting electronic devices and other gadgets. The study of supercap batteries is on the rise to obtain high specific power and energy density along with cycle stability. Chapter 6, "IPMC Based Flexible Platform: A Boon to the Alternative Energy Solution," presents a possible solution to the problems of ionic polymer-metal composite membranes. The ionic polymer–metal composite-based platform was seen as a boon to the alternative energy solution. The presented study examines energy storage applications and the practical characteristics of ionic polymer-metal composite polymer materials in the analogous field. Chapter 7, "Structural, Optical, and Electrical Studies of PAN-Based Gel Polymer Electrolytes for Solid-State Battery Applications," is a study of a polyacrylonitrile (PAN) gel polymer electrolyte system for solid-state battery applications. in recent times, ionic solid conducting polymer electrolytes have been used for active potential components in novel battery technology as electrodes and electrolytes. The results obtained in the study were found to be comparable with existing solutions. Finally, Chapter 8, "Phase Change Materials for Renewable Energy Storage Applications," is of key interest because super thermal properties of advanced materials are required to enhance maximum utilization of solar energy and improve energy and exergy efficiency of the solar absorbing system. Phase change materials have high energy density and potential in flat plate solar collector application for the production of hot water in urban households. Thus, they could be successfully incorporated in integrated and non-integrated flat plate solar collectors.

This book examines new trends of research into energy storage devices as applied in renewable energy and smart grid applications. It also addresses shortcomings in the implementation of energy storage. This volume is a useful reference for readers who wish to familiarize themselves with the newest advancements in energy storage systems.

**Dr. Kenneth E. Okedu**
Visiting Professor,
Department of Electrical and Communication Engineering,
National University of Science and Technology,
Muscat, Sultanate of Oman

Adjunct Professor,
Department of Electrical and Electronic Engineering,
Nisantasi University,
Istanbul, Turkey

Section 1

# Batteries Storage Techniques

Chapter 1

# Electrical Equivalent Circuit Models of Lithium-ion Battery

*Rushali R. Thakkar*

## Abstract

Modelling helps us to understand the battery behaviour that will help to improve the system performance and increase the system efficiency. Battery can be modelled to describe the V-I Characteristics, charging status and battery's capacity. It is therefore necessary to create an exact electrical equivalent model that will help to determine the battery efficiency. There are different electrical models which will be discussed and examined along with the benefits and demerits. A systematic comparison and analysis using simulation will help us to select an ideal model which will suit best to a specific application.

**Keywords:** battery, lithium-ion batteries, battery modelling, state of charge, electrical equivalent, simulation

## 1. Introduction

Renewable energies play a major role in the power quality applications of emerging technologies. Batteries are considered as one of the most important amongst the component. Batteries are energy storage devices that can be utilised in a variety of applications and range in power from low to high. Batteries are connected in series and parallel to match the load requirements. The advantages of lithium-ion batteries include their light weight, high energy density, and low discharge rates. They're commonly seen in high-power applications like electric vehicles and hybrids. Lithium-ion batteries have a terminal voltage of 3-4.2 volts and can be wired in series or parallel to satisfy the power and energy demands of high-power applications. Battery models are important because they predict battery performance in a system, designing the battery pack and also help anticipate the efficiency of a system [1, 2].

## 2. Existing electrical equivalent battery models

The mathematical relationship between the elements of Lithium-ion batteries and their V-I characteristics, state of charge (SOC), internal resistance, operating cycles, and self-discharge is depicted in a Lithium-ion battery model. The equivalent circuit model of a Lithium-ion battery is a performance model that uses one or more parallel combinations of resistance, capacitance, and other circuit components to construct an electric circuit to replicate the dynamic properties of Lithium-ion batteries. Time domain analysis is used to produce the most often utilised electrical equivalent models. The simplest model equation for battery model can be represented by Open Circuit Voltage (OCV)

$$v(t) = OCV \tag{1}$$

SO0C of a cell is 100% when cell is fully charged and SOC is 0% when cell is fully discharged. The amount of charge removed from 100–0% is the total capacity measured in Ah or mAh.

Following are the models used for Lithium-ion battery to determine the performance on the system [3].

## 2.1 Simple battery model Rint model

**Figure 1** illustrates the most widely used battery model. It consists of an ideal battery with open circuit voltage (OCV) $V_O$, a constant equivalent internal resistance $R_{int}$, and $v(t)$ as the terminal voltage. When completely charged, the terminal voltage v(t) may be determined by measuring the open circuit voltage, and Rint can be determined by connecting a load and detecting both the terminal voltage and current.

SOC is represented as z(t) of a cell is 100% when cell is fully charged and SOC is 0% when cell is fully discharged. The amount of charge removed from 100–0% is the total0 capacity (Q) measured in Ah or mAh.

$$v(t) = OCV(z(t)) - i(t)R_{\text{int}} \tag{2}$$

$$\frac{dz}{dt} = -\frac{i(t)}{Q} \tag{3}$$

$$z(t) = z(t0) - Q\int_{t0}^{t} i(T)dT \tag{4}$$

At the point when a battery is completely charged, its Open Circuit Voltage is higher than when it is discharged. The Rint model appears to be relatively simple, however it ignores the variable nature of the internal resistance related to temperature, state of charge, and electrolytic concentration.

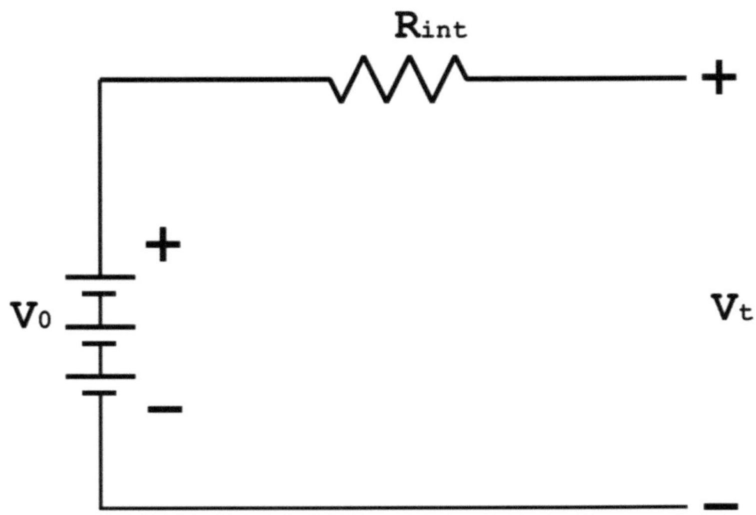


**Figure 1.**
*Resistive Thevenin battery model.*

### 2.2 Resistive Thevenin battery model

**Figure 2** shows the circuit schematic for the resistive Thevenin battery model. This model has two types of internal resistances, $R_0$ and $R_1$, which are connected to the charging and discharging properties of the battery, respectively. Electrical and non-electrical losses are modelled by the internal resistances $R_0$ and $R_1$. Diffusion Voltages also can be closely found using1 or more RC parallel branches. The model gives better results than $R_{int}$ model, however transient conditions like the capacitance effect are not considered. As a result, this model is non-dynamic and unsuitable for applications involving electric vehicles and electric hybrids [4–8].

In relaxation mode, voltage gradually reduces to zero, this is termed as diffusion voltage which can be closely approximated using parallel RC branches (**Figure 3**) [4, 5].

$$v(t) = OCV(z(t)) - i(t)Rint - Vc1(t) \tag{5}$$

### 2.3 An accurate electrical battery model

An Accurate Electrical Battery Model, models the battery capacity, charging state, and run time using a capacitor and a current controlled source. The circuit takes into account the battery life time as well as the slow and fast transient response. A voltage controlled voltage source that is a function of state of charge is used to overcome the barrier between SOC and OCV (**Figure 4**) [1].

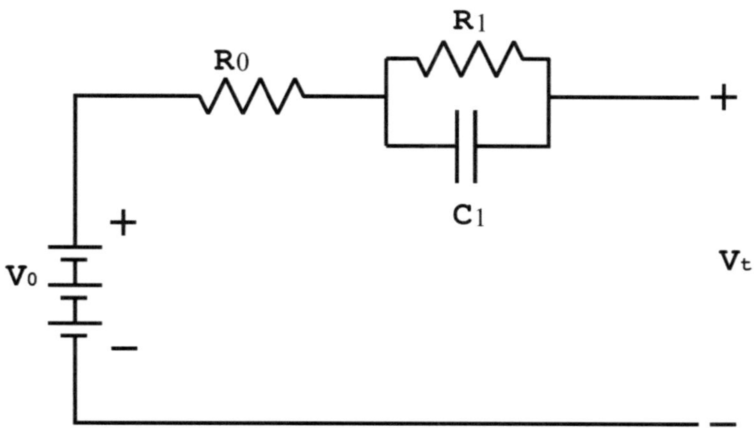


**Figure 2.**
*Thevenin battery model with 1 RC branch.*

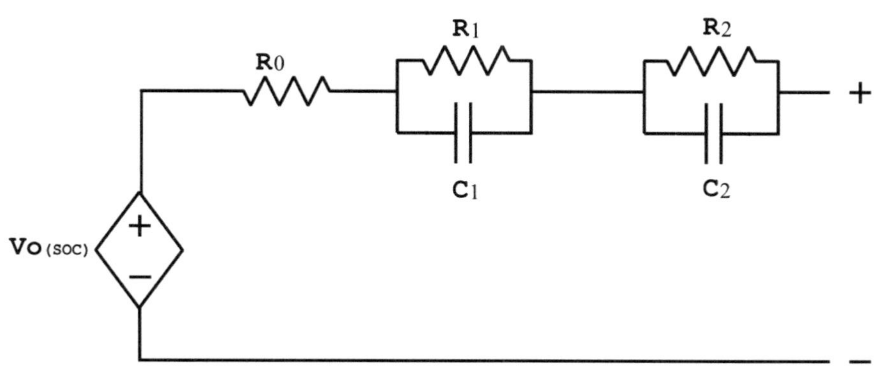


**Figure 3.**
*Thevenin battery model with 2 RC branches.*

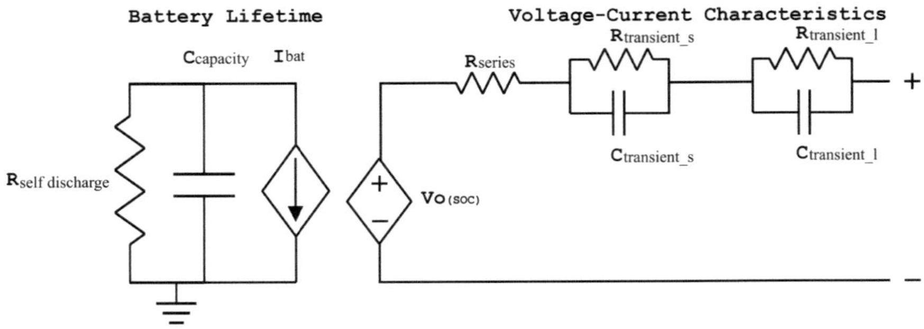


**Figure 4.**
*Accurate electrical battery model.*

| S. No. | Battery model | Features |
|---|---|---|
| 1 | $R_{int}$ model | Model is static. |
| 2 | Thevenin electric model | Does not consider the dependency of state of charge. |
| 3 | Second order RC Thevenin model | Model does not consider the parameters like temperature and capacity. |
| 4 | Accurate electrical equivalent model | Model considers the battery life time. |

**Table 1.**
*Comparison of electrical equivalent battery models [8].*

| Capacity | Nominal capacity: 2600 mAH (0.52 A discharge, 2.75 V) |
|---|---|
| Nominal voltage | 3.7 V |
| Internal impedance | ≤70 mΩ |
| Discharge cut off voltage | 3.0 V |
| Maximum charge voltage | 4.20 ± 0.05 V |
| Standard charge discharge current | 0.52 A |
| Rapid charge discharge current | 1.3 A |
| Maximum pulse discharge current | 2.6 A |

**Table 2.**
*Battery model specifications LIR18650 mAH.*

## 2.4 Models of electrical equivalent batteries

**Table 1** gives the different Electrical equivalent circuit models along with its important features so that it will be useful to predict the performance of the battery and for which application the battery can be used for.

## 2.5 Electrical characteristics of lithium-ion battery

Lithium-ion battery specifications used for battery model: LIR18650 mAH are given in the following **Table 2** [8].

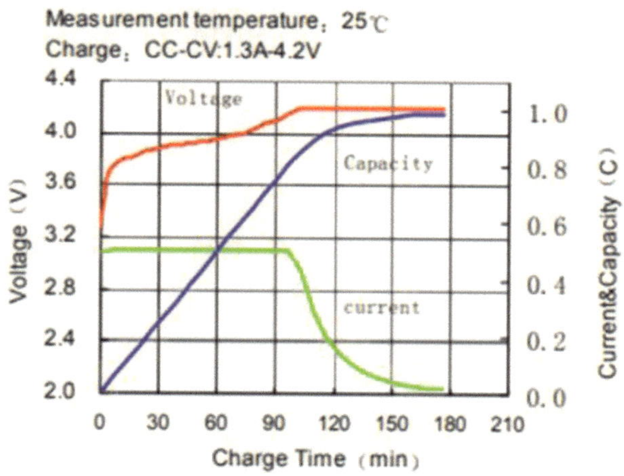


**Figure 5.**
*Rate charge characteristics.*

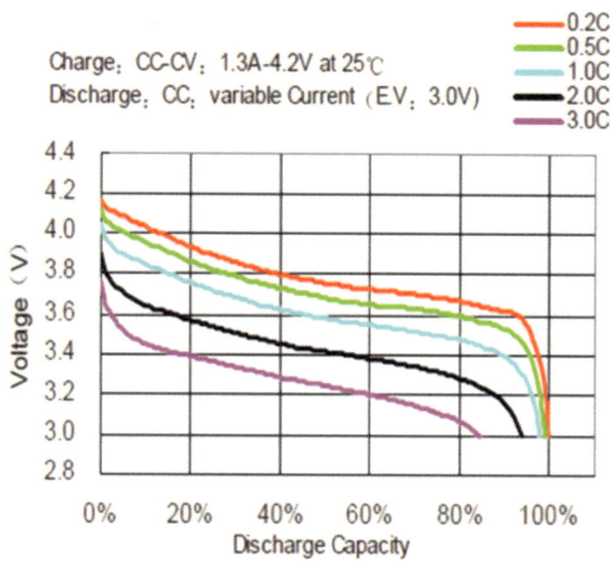


**Figure 6.**
*Rate discharge characteristics.*

## 2.6 Charging and discharging characteristics

**Figures 5** and **6** shows the rate charge and discharge characteristics of the battery Model LIR18650 2600 mAH. The battery charges with Constant Current Constant Voltage mode. The battery is charged at a constant current until it reaches 4.2 volts, then it is charged at a constant voltage until the current drops to zero [8]. The charging time of battery can be calculated as the ratio of battery Ah to the charging current.

# 3. Simulation results and discussion

M file is created in Matlab to simulate the model's charging and discharging curves. The generalised model for lithium-ion batteries uses the equations below [7, 8].

Discharge Model (i*>0)

$$f1(it, i*, i) = E0 - K.\frac{Q}{Q - it}.i* - K.\frac{Q}{Q - it} + A.\exp(-B.it) \quad (6)$$

Charge Model (i*<0)

$$f1(it, i*, i) = E0 - K.\frac{Q}{it + 0.1.Q}.i* - K.\frac{Q}{Q - it} + A.\exp(-B.it); \quad (7)$$

E0 is constant voltage (V), K is polarisation constant in ($Ah^{-1}$), i* is low frequency current dynamics, Q is maximum battery capacity (Ah), A is exponential voltage (V), B is exponential capacity ($Ah^{-1}$), *it* is extracted capacity (Ah).

The implementation of the generalised model in MATLAB shows that the characteristics developed by mathematical electrical model of battery in Matlab and are close to the actual characteristics of Lithium-ion Battery Model LIR18650 2600 mAH [8].

## 4. Conclusion

Being a static model, the Rint model does not account for the battery's charging and discharging properties. Thevenin's electric model (1RC) ignores dynamic behaviour and neglects to account for state of charge dependency. 2 RC branches would be closer to match the data at the beginning of the transient. 3RC branches can be chosen as a compromise between complexity of model and its fidelity. Various electrical equivalent models are studied along with its different features in the existing models.


## Author details

Rushali R. Thakkar
Sardar Patel Institute of Technology, Mumbai, India

*Address all correspondence to: rushalithakkar@gmail.com

## References

[1] Min Chen and Gabriel A. Rinc'on-Mora. Accurate electrical battery model capable of predicting runtime and I–V performance. IEEE Transactions on Energy Conversion 2006;21(2).

[2] Xiaoqiang Zhang, Weiping Zhang, and Geyang Lei. A review of Li-ion battery equivalent circuit models. Transactions on Electrical and Electronic Materials 2016;17(6):311-316.

[3] Low Wen Yao, J. A. Aziz, Pui Yee Kong, and N. R. N. Idris. Modeling of lithium-ion battery using MATLAB/Simulink. In: Proceedings of Conference of the IEEE Industrial Electronics Society. Vienna, pp. 1729-1734, Nov. 2016.

[4] Barcellona Simone and Piegari Luig. Lithium ion battery models and parameter identification techniques. Energies 2017;10.

[5] Zhang Lijun, Peng Hui, Ning Zhansheng, Zhongqiang Mu, and Sun Changyan. Comparative research on RC equivalent circuit models for lithium-ion batteries of electric vehicles. Applied Sciences 2017;7.

[6] Hyun You, Jun Bae, So Cho, Jong Lee, and Se-Hun Kim. Analysis of equivalent circuit models in lithium-ion batteries. AIP Advances 2018;8.

[7] Haizhou Zhai. Modeling of lithium-ion battery for charging/discharging characteristics based on circuit model. International Journal of Online Engineering, 2017;13(6):86-95.

[8] R. R. Thakkar, Y. S. Rao, and R. R. Sawant. Performance Analysis of electrical equivalent circuit models of lithium-ion battery. In: 2020 IEEE Pune Section International Conference (PuneCon), 2020, pp. 103-107. DOI: 10.1109PuneCon50868.2020.9362386

Chapter 2

# Battery Energy Storage Systems and Rooftop Solar-Photovoltaics in Electric Power Distribution Networks

*Innocent E. Davidson and Rodney Reddy*

## Abstract

Energy storage technologies is transforming the way the world and utility companies utilize, control and dispatch electrical energy. In several countries, the consequential effect of meeting electrical demands continues to burden the electrical infrastructure leading to violation of statutory operating limits. Such violations constrain a power system's ability to supply suitable energy whilst meeting daily load and growth demands. While optimization techniques can be used to reduce violations, these are still limited do not provide effective short-term solutions when dealing with constrained networks in dense and radial distribution systems. Battery energy storage systems (BESS) and solar rooftop photovoltaics (RTPV) are a viable distributed energy resource to alleviate violations which are constraining medium voltage (MV) networks.

**Keywords:** battery energy storage systems (BESS), rooftop solar-photovoltaics (RTPV), power distribution, voltage limits, thermal limits, technical performance

## 1. Introduction

In designing and operating electric power networks or implementing major expansions to existing networks, a number of key issues regarding the technical performance of the network at both the transmission and distribution (T and D) levels must be ascertained. These include voltage regulation, voltage fluctuations, rapid voltage rise, electrical loses, distribution plant loading and utilization, fault level, generation stability, harmonics, phase balancing, system security, and supply availability [1]. The approach of Power Utilities to address any constraints or violations, experienced within medium voltage networks, would be based on extensive technical evaluation. While overvoltage is a concern if roof-top solar-photovoltaic (RTPV) penetration is not regulated [2], this study shows the benefit of RTPV and/or including battery energy storage systems (BESS), as this offers relief for constrained networks.

## 2. Network model selection and appraisal

Real-time power system analysis deals with two critical criteria, from the network appraisal, this being Voltage and Thermal constraints/violations. From the analysis, a network was identified which has both voltage and thermal violations. This study explores the technical influence that RTPV has on MV networks. From the appraisal analysis, a most fitting network type was identified which is found most common amongst the feeders that were analyzed. The predominate network classification is found to be a C2, TZ2 type network [3, 4]. In-line with the investigative violations (voltage and thermal) these networks have a tolerable minimum voltage of 95.5% during normal conditions, and 93.5% during abnormal conditions.

The thermal limit defined in **Table 1** indicates this to be alarming above 90% of its rated capacity.

The MV network selected for this study is Madadeni NB36. **Table 2** shows the appraisal results of Madadeni NB36; this network has been selected due to having both voltage and thermal violations. This study goes on to showcase how RTPV and/or BESS influences the violations identified on NB36.

As stated in **Table 2**, NB36 peaks during a winter's weekday at 18:30. Annual statistical metering data was analyzed and using statistical analysis methods, data has been fashioned into four, twenty-four hours, thirty-minute intervals; seasonal profiles, based on the four weather seasons. The load profile for a winter's weekday demonstrates the peak loading period and is used throughout this chapter to demonstrate the effective influence of RTPV and/or BESS.

**Figure 1** shows a typically generated PV profile which have been defined to represent each season [7, 8]. The duration of sunlight hours differs between each season. To be consistent with the network analysis, the winter PV profile was selected to match the network breakers winters statistical load profile.

When considering RTPV, some configurations should be considered to decide on the most cost-effective or technically beneficial solution for both the Utility and Customer. This chapter considers the following configurations:

- BESS alone
- RTPV alone
- RTPV with BESS
- Varied levels of BESS

Power simulation has been conducted in Power Factory, the results have been extracted, analyzed and stated in the legends field within the applicable figures.

| Network type | Feeder type | Network class | Network voltage (kV) | Conductor size/type | Maximum allowed MV across backbone | Maximum allowed MV line/ cable loading |
|---|---|---|---|---|---|---|
| Urban | Cable | C1 | 11 | 185 $mm^2$ | 2% | 50% of normal to allow for N-1 |
| Urban | Cable | C1 | 22 | 95 $mm^2$ | 2% | 50% of normal to allow for N-1 |
| Rural | Overhead | C2 | 11 | Hare | 7.5% | 90% of rate A[3] |
| Rural | Overhead | C2 | 22 | Hare | 7.5% | 90% of rate A |

**Table 1.**
*Network characteristics for calculating customer number limits [5].*

| Parameter | Value | Units |
|---|---|---|
| Nominal Voltage | 11 | kV |
| Time of Feeder Peak | Winter-WD-18:30 | Seasonal Time |
| Feeder Peak S | 4.35 | MVA |
| Feeder P at Peak S | 4.23 | MW |
| Feeder Q at Peak S | 0.58 | MVARs |
| Feeder PF at Peak S | 0.97 | pf |
| OC Setting | 250 | A |
| Load % of OC Setting | 93 | % |
| Feeder Minimum Voltage p.u. | 92.7 | % |
| Max Equipment Loading % | 102 | % |
| MV/LV (11-22 kV/400-230 V) Xfr Count | 65 | Count |
| Installed Capacity | 5.29 | MVA |
| Total Line Length | 23.77 | km |
| Backbone Length | 13.79 | km |
| Customer Base | 3378 | Count |

**Table 2.**
*Madadeni NB36 appraisal data [6].*

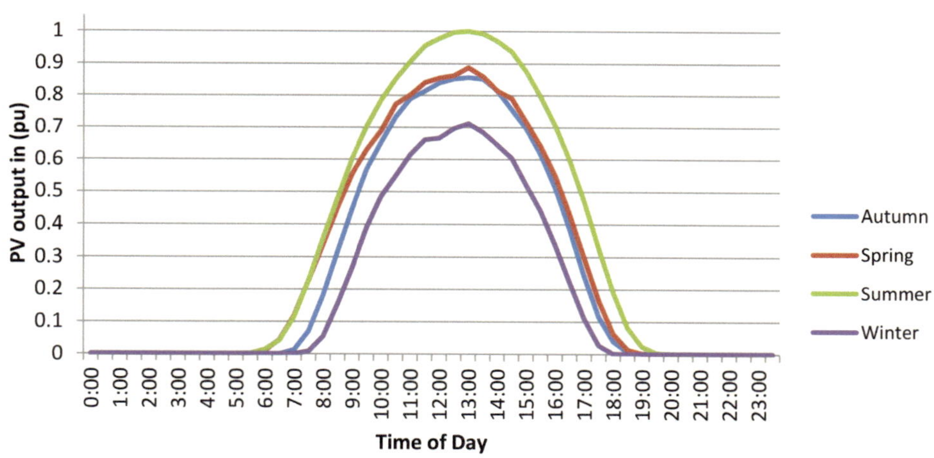


**Figure 1.**
*Typical p.u. generated seasonal PV output profile.*

## 3. Installed RTPV system

The scope of this study relates to RTPV installations connected to every customer connected to an MV network, this is defined as the high penetration of RTPV. The equipment type used in this study is found to be commonly available in South Africa.

While some studies consider large PV installation's sizes to show benefits to customers [9], this study utilizes an average solar panel output of a single 200 W panel installed at every connected customer [10], this is a very conservative approach taking into account that panel outputs degrade over time due to aging and associated output reduction [11], the build-up of dirt/residue and orientation.

Further to this the battery technology assumed in the assessments, also commonly available, are lead-acid type batteries. These batteries have a depth of discharge rate (DOD) and while it is common to find a DOD of 80%, for the purposes of this study it is assumed that a 100-ampere hour battery with a DOD or 50% is utilized offering 600 W of power. Inverter capabilities [12] have been assumed to operate at 7 amps due to low household breaker sizing and inverter costings. This simulation design, though very conservative, leads to defining the required design for households. The criteria of the PV, battery and inverter that have been utilized in this study are considered as a base design which can be improved upon implementation.

## 4. Battery energy storage system alone

Considering battery energy storage systems (BESS) without any external generating source would require that its charging is supported by the electrical grid. Analysis in [13] shows that there is not any reduction in the total daily energy supplied by the Utility, hence the only benefit that can be achieved would be for the Utilities' peak load shaving. This arrangement shows no customer benefit and does not result in any energy saving for the Power Utility either. For a consumer with a BESS installation only, a likely benefit is to run on charge cycles during periods of low load and inject at periods of high utility demand especially if tariffs are based on time of use.

## 5. Roof-top photovoltaic alone

Analysis has been considered for each residential customer having 200 W (watts) of installed roof-top photovoltaic (RTPV) capacity, an overview of this connection can be seen in **Figure 2**. It is assumed that if the power generated by the RTPV would exceed their instantaneous demand then the excess power would feed into the electrical grid.

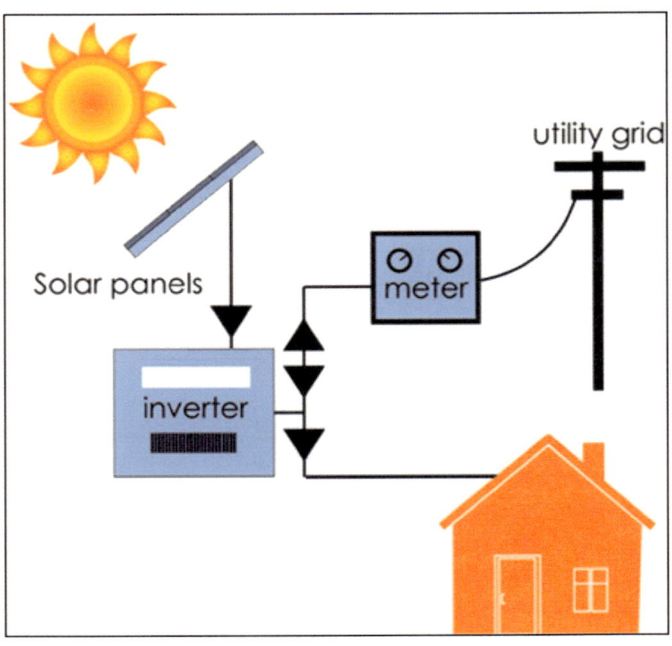


**Figure 2.**
*Conventional RTPV installation [14].*

**Figure 3** shows the comparison of RTPV and the network normal power curves. The network normal (P) is the load profile for the study. It can be seen that during daylight hours RTPV dispatches power; hence the reduction of power can be seen when compared to the network normal curve.

From this analysis, it is seen that the Utility loses 5.76% in revenue resulting from an overall reduction in dispatched power. While the supplied power to the customer from the utility is reduced by 5.8%, referring to data in **Figure 4**. Hence the

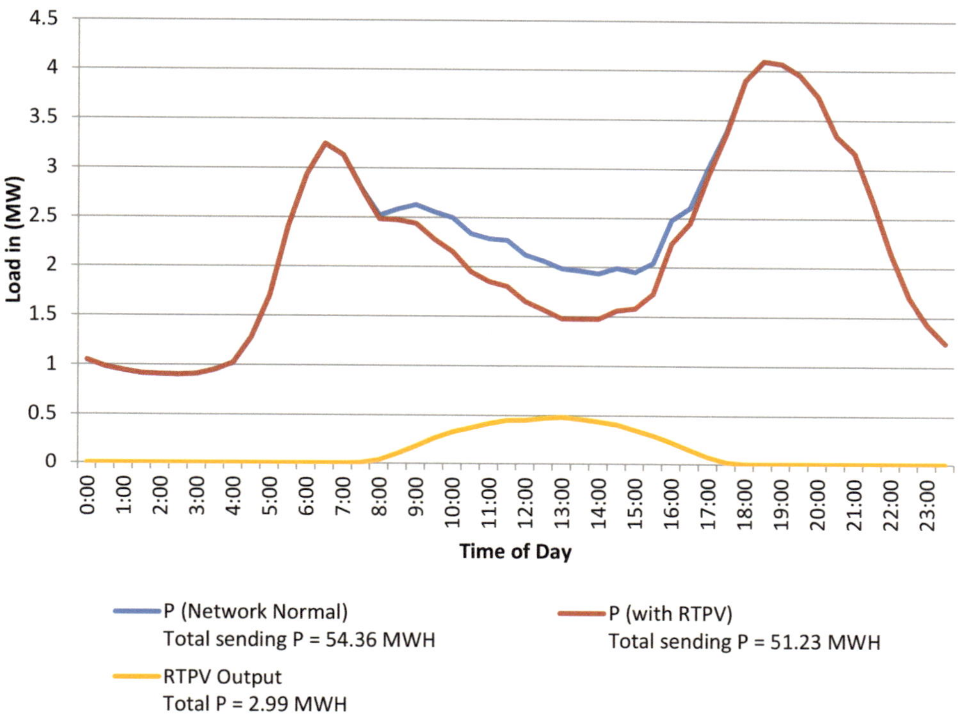


**Figure 3.**
*Active power of NB36 with 200 W RTPV penetration, considered for a winter weekday.*

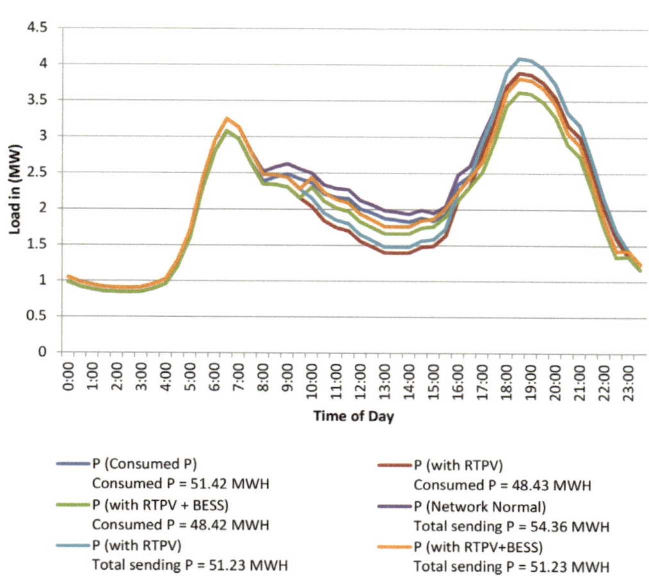


**Figure 4.**
*Sending active power, power consumed by customers and P losses, considered for a winter weekday.*

benefit to the customer is credited to the RTPV injection. The disadvantage when considering RTPV alone is that it feeds only during daylight hours, there is no effect to the network morning and evening peaks period; the peak violations mainly exist during the evening peak time. This leaves the question of what will be the influence of RTPV combined with BESS.

## 6. Roof-top solar photovoltaic with battery energy storage system

Considering the same RTPV installed capacity of 200 W per residential home. In addition to this, it is assumed that each home is equipped with a battery which has 600 W of dispatchable power; an overview of this connection is seen in **Figure 5**. This inclusion of BESS is limited only by its charge and discharge rate. Based on the available power generated by a 200 W RTPV source, it was then assumed that the battery would reasonably charge and discharge at less than 7 amps.

The network normal (P) is the load profile for the study. **Figure 6** shows the comparison of RTPV with and without BESS. During daylight hours 80 W of power is allowed to charge the battery while the remaining power supplies the customer or overflows onto the electrical grid. During the evening peak period, the same 80 W is dispatched from the stored energy over a period of 7.5 hours.

As demonstrated in **Figure 6**, RTPV including BESS can be seen to benefit both the Utility and Customer. The Utility benefits by the reduced voltage regulation and thermal violations which are experienced during the evening peak, and the

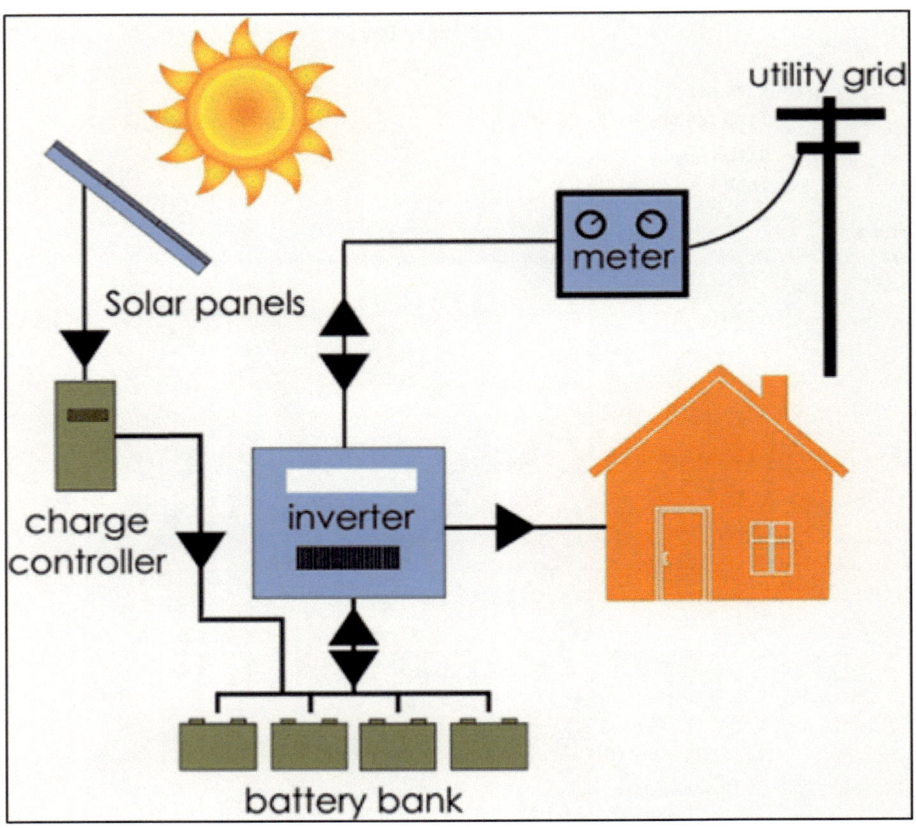


**Figure 5.**
*RTPV with battery storage [15].*

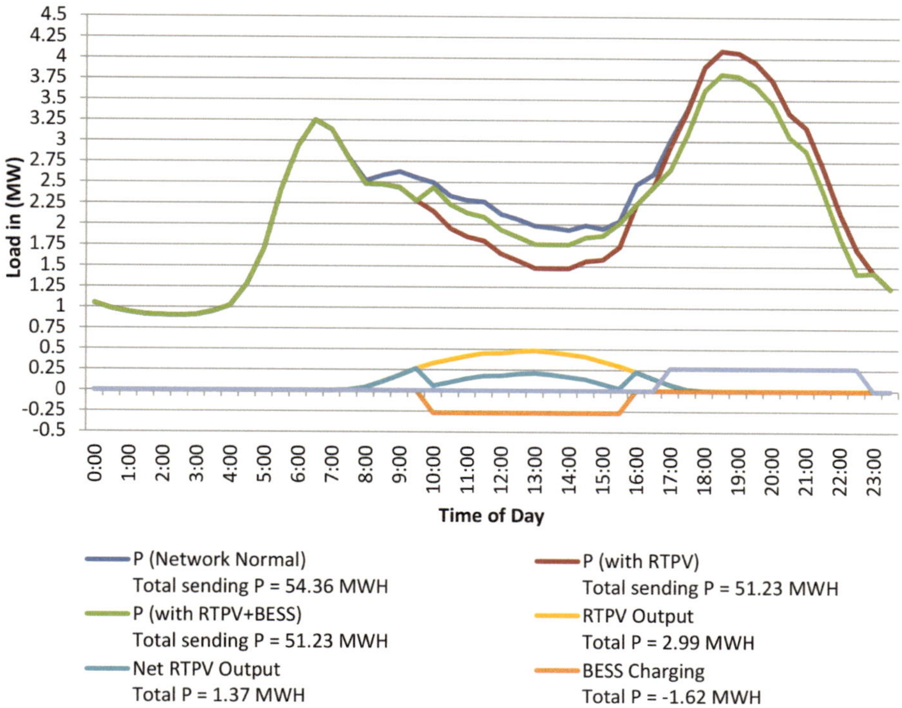


**Figure 6.**
*Active power of NB36 with 200 W RTPV penetration and BESS, considered for a winter weekday.*

Customer benefits by an overall reduction in power consumed. From this analysis, it is seen that the Utility loses 5.76% in revenue. While the customer's power consumption, referring to data in **Figure 4**, is reduced by 5.8%. Further investigation of **Figure 4** shows a negligible difference of power consumed if the customer has RTPV alone or including BESS.

## 7. Varied levels of BESS

While the results from the above analysis speak of how the Utility and Customers are affected, there still raises the fundamental concern related to the constraints experienced on NB36. At normal, when the load is peaking, it's found that the minimum Voltage is 92.7% and the network is thermally loaded to 102%. The analysis shows that if customers provide back-feeding, from their stored energy, during peak times, it will reduce the constraints on the network. Analysis has considered the back-feeding of 80 W, following the discussion above, and 250 W for each residential customer.

**Figure 7** reflects the effect of the voltage along the backbone of NB36, also showing the ability of BESS to dispatch power at 80 W and 250 W. It can be seen that with an increasing ability of the BESS to dispatch power, it alleviates the voltage constraint by <1% at 80 W, and 2.1% at 250 W.

**Figure 8** shows that the normal network condition exceeds the rating of the conductor at peak loading.

With BESS dispatching power we see that the thermal constraint is reduced from 102–94% when 80 W of power is dispatched, and improved further when 250 W is dispatched to 79.8%.

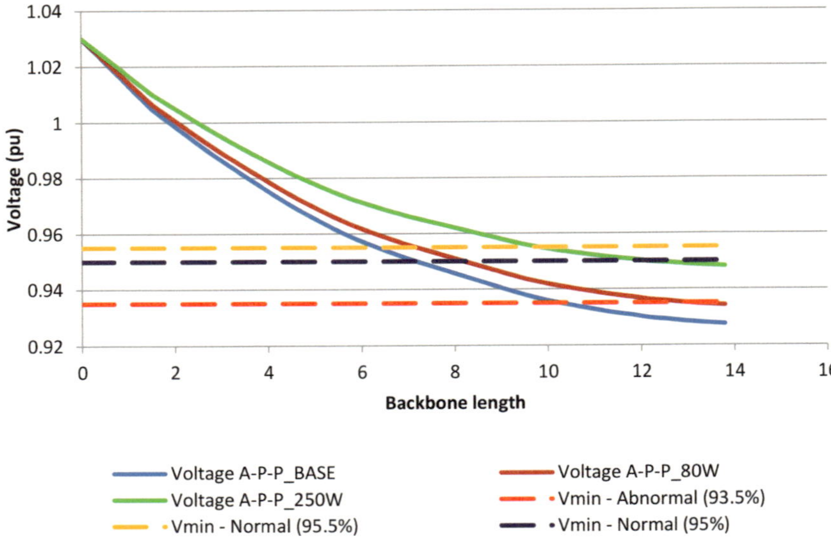


**Figure 7.**
*NB36 minimum voltage at peak (winter weekday – 18:30).*

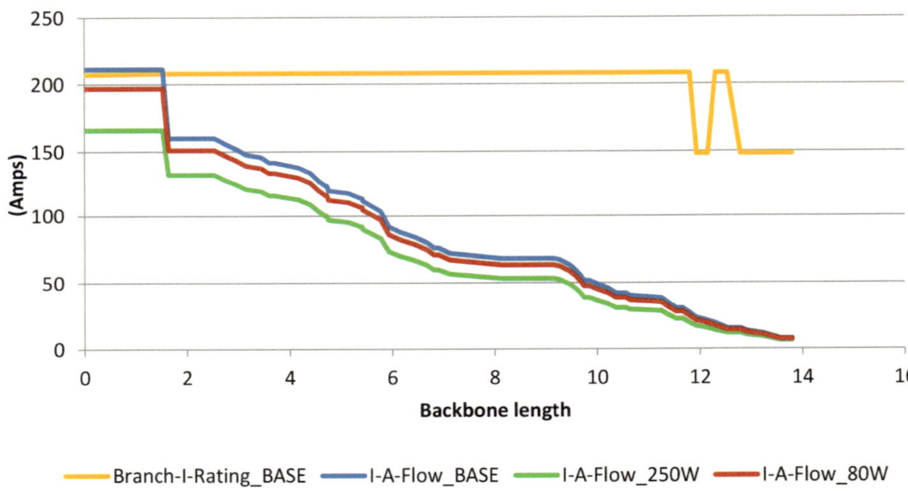


**Figure 8.**
*NB36 thermal rating vs. load current.*

## 8. Power dispatched, consumed and its effect on technical losses

**Figure 9** shows the network sending power, consumed power (including with RTPV, and RTPV+BESS). The difference between the sending power and consumed power is attributed to the technical losses resulting from transmitting power down the network.

Technical losses are calculated as a percentage of the total load of the feeder including both no-load and load losses. The results from the analysis help to obtain a view of the network losses; this can be used as a basis for historic trending and benchmarking and can be used as one of the triggers for network strengthening. Therefore, this statistic also aids the conditioning of the priority ranking criteria. DER's may have a significant effect on network losses.

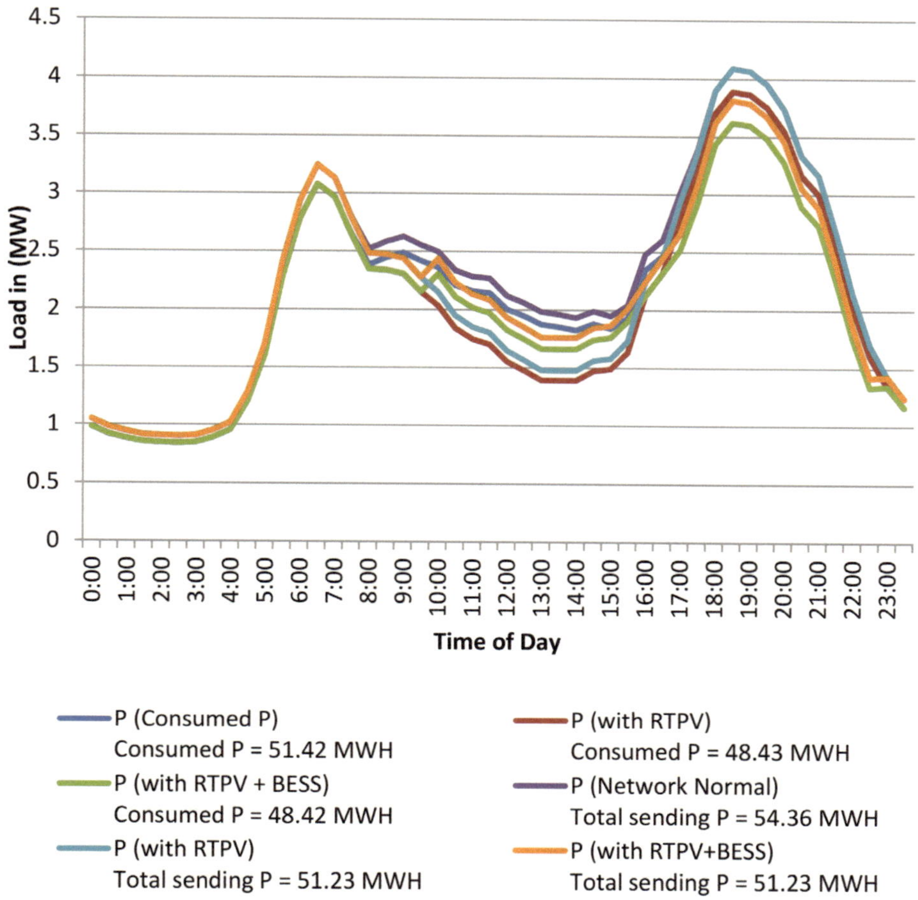


**Figure 9.**
*Sending active power, power consumed by customers and P losses, considered for a winter weekday.*

A generator can lower or increase losses, depending on its location and the network configuration [16].

The technical losses vary slightly between the different options analyzed:

- Network sending power vs. Consumed power (Network normal) = 5.41%
- Network sending power vs. Consumed power (with RTPV) = 5.47%
- Network sending power vs. Consumed power (With RTPV+BESS) = 5.49%

## 9. Conclusion

This chapter has demonstrated the benefit of roof-top solar photovoltaic and/or including battery energy storage systems. It offers relief for constrained networks in dense and radial distribution systems. While optimization techniques can be used to reduce violations, these are still limited do not provide effective short-term solutions when dealing with constrained networks in dense and radial distribution systems. Battery energy storage systems (BESS) and solar rooftop photovoltaics (RTPV) are a viable distributed energy resource to alleviate violations which are constraining medium voltage (MV) networks. The results show the following:

### 9.1 BESS only

This option does not benefit the customer, as the batteries require grid connection to charge and discharge, not to mention efficiency losses. If discharging the stored energy occurs during peak periods, this can benefit the utility by reducing the peak violations. Therefore, it is recommended that a tariff/time-of-use incentive is introduced to motivate customers for BESS only installations.

### 9.2 RTPV only

This is an excellent way for a customer to reduce his overall electrical utility bill. Unfortunately, the utility is negatively affected by the reduced sale of electricity.

### 9.3 RTPV+BESS

The results show that with the addition of RTPV including BESS the utility still loses revenue. However, with the addition of BESS, Utilities have the ability to reduce technical violations during peak periods. Installing BESS, in this manner, shows no benefit to the customer. Therefore, it is recommended that a tariff/time-of-use incentive is introduced to motivate customers for RTPV+BESS installations.

### 9.4 Varied levels of BESS

Though RTPV inclusive of BESS reduces the violations on the network; it can be seen from **Figures** 7 and **8** that voltage and thermal violations persist for this specific network. Therefore, it is necessary to consider an increased installation of BESS. 250 W of dispatchable power offered the most appropriate quantity of power to alleviate the violations. The effect on technical power losses: seen in **Figure 9**, shows a constant 5.41%–5.49% of technical power losses for each of the above considerations. This demonstrates that technical losses are similarly proportioned to its source sending power when conducting analysis on the various installation types relating to RTPV/BESS. While these amounts are typical and expected for a reticulation network – losses will defer due to topology, loading and design of networks.

For Utilities and Municipalities, the extent of challenges encountered when considering large scale installations of RTPV would be related to availability and visibility of data for adequate analysis. Current standards do not address the practical design solutions needed for all variations within customer installations and expectancy from RTPV. This can result in non-standard customer installations which will lead to undesirable impacts on the source and shared utility power systems.

Visibility and compliance of electricity supply regulations are required at the point of supply which is conditioned to be met at the point of supply and not the point of generator connection which is embedded in the customer's installation. Therefore, smart metering systems with predefined charge and discharge times/durations need to accompany RTPV installations. This requires an evaluation of currently employed metering technology, to ensure they accommodate these operational scenarios.

The cost for an electrical system, in which a customer is partially or completely off-grid, will still be attributed to the Utility to ensure the security of supply. Hence Utilities should take the lead in this segment of small-scale embedded generation to remain viable and relevant. Utilities may consider engaging suppliers to carry out maintenance and/or repairs during initial warranty periods.

Data visibility, both topological and metered data, is crucial for power system analysis, and need to be available and validated from time to time to cater for load growth. Utilities must ensure competent staffing for data acquisition and analysis.

There may be a lack of data validity for non-telemetered devices, such as unknown tap positioning of reticulation transformers, conductors or cables which has been replaced with different ratings and types. This can be challenging when configuring simulation models.

For this study analysis was given to the network classification that was predominately found with residential type loads, which has been identified as C2 TZ2 type. This study, therefore, adopted a common sending voltage regulation set point of 1.03pu applied at reticulation busses. The voltage limit allowable for a normal condition is 95.5% and for abnormal conditions are 93.5. From an operations view, the abnormal limit is the benchmark to be adhered to, while network planners work with the normal limit of 95.5%.

Continuous network changes due to rising electrification and illegal connections discredits network analysis and requires more frequent update of simulation models. Pre-existing RTPV installations may pose a problem if they do not meet compliance requirements. Utilities need to conduct surveys of pre-existing RTPV installations and update its electrical connection to the power grid.

It is recommended that during site visits, meter re-programming including 'time-of-use' tariffs and amendments to supply agreements, can be implemented. Presently, there is no information on households which has become "self-suppliers" due to lack of a registry for small-scale embedded PV plants. This will assist in ensuring that customer contributions to unbalance be limited to 1% voltage unbalance at the point-of-common coupling.

## Acknowledgements

Thanks to Dayahalen Chetty of Eskom Holdings SOC Ltd.,

## Author details

Innocent E. Davidson* and Rodney Reddy
Durban University of Technology, Durban, South Africa

*Address all correspondence to: innocentd@dut.ac.za

## References

[1] Ogunboyo PT, Tiako R, Davidson IE, "Effectiveness of dynamic voltage restorer for unbalanced voltage mitigation and voltage profile improvement in secondary distribution system," in Canadian Journal of Electrical and Computer Engineering, vol. 42, no. 2, pp. 105-115, Spring 2018. doi:10.1109/CJECE.2018.2858841.

[2] Goqo Z, Davidson IE, "A Review of Grid Tied PV Generation on LV Distribution Networks". Proceedings of the IEEE Power Africa Conference, 26-29 June, 2018, Cape Town, South Africa, pp. 310-315.

[3] Bello M, Chetty D. "Distribution Network Operations Planning (DNOP) Standard. 240-82534300." Eskom Holdings Ltd, Feb. 2015.

[4] Bello M, Brown CC. "Distribution Voltage Regulation and Apportionment Limits, DST_34-542." Eskom Holdings Ltd, Jul. 2014.

[5] Kleynhans T, Gutschow D, Merwe JVD. "Planning Standard for Distribution Network Reliability to Ensure Distribution Network Code Compliance. 240-76613395." Eskom Holdings Ltd, 2015.

[6] Global Solar Atlas, 'Madadeni NB36 PV Power Output Report for 0.200kWp installation, Generated by Global Solar Atlas'. [Online]. Available: https://globalsolaratlas.info. [Ref A2]

[7] Global Solar Atlas, 'Madadeni NB36 PV Power Output Report for 0.200kWp installation, Generated by Global Solar Atlas'. [Online]. Available: https://globalsolaratlas.info.

[8] Kern Jr EC, Russell MC. "Experiences and Lessons Learned with Residential Photovoltaic Systems. No. EPRI-GS-7227." Electric Power Research Inst., Palo Alto, CA (United States); Ascension Technology, Lincoln Center, MA (United States); Southwest Technology Development Inst., Las Cruces, NM (United States), 1991.

[9] Niemz B. "Can low-income homes be made to benefit from the energy transition?", Creamer Media's Engineering News, vol. 39 no 16, pp. 60-61, May 03, 2019.

[10] Vasili A. "Solar Panel Output: How Much Electricity Do They Produce?" The Eco Experts, May 26, 2020. [Online]. www.theecoexperts.co.uk/solar-panels/electricity-power-output.

[11] Jordan DC, Deline C, Kurtz SR, Kimball GM, Anderson M. "Robust PV degradation methodology and application." IEEE Journal of Photovoltaics 8, no. 2 (2017): 525-531.

[12] Koutroulis E, Blaabjerg F. "Design optimization of transformerless grid-connected PV inverters including reliability." IEEE Transactions on Power Electronics 28, no. 1 (2012): 325-335.

[13] Thopil M, Moodley GV, Jennings GD, Buyisa D. "Technical and financial impacts of residential PV-battery systems." In Proceedings of the 26th AMEU Convention, pp. 90-94. 2018.

[14] "Grid-Tied System (No Battery Back-Up)." Solarent, Oct. 2020. [Online]. www.solarent.com/about-solar/photovoltaic-solar-panels.php.

[15] "Grid-Tied/Hybrid System with Battery Storage." Solarent, Oct. 2020. [Online]. www.solarent.com/about-solar/photovoltaic-solar-panels.php.

[16] Bello M. "Network & Grid Planning Standard for Generation Grid Connection – Generators Technology Overview and Effects on Networks, 34-1944." Eskom Holdings Ltd, Jul. 2014.

Chapter 3

# The Second Life of Hybrid Electric Vehicles Batteries Methodology of Implementation in Ecuador

*Efrén Fernández Palomeque, Diego Rojas Hiedra, Daniel Cordero and Martín Espinoza*

## Abstract

Hybrid car sales in Ecuador in the last 10 years are very promising. The presence of hybrid electric vehicles (HEV) in the country generates an increase in nickel metal hydride batteries used (NiHm), these batteries do not follow an adequate recycling and disposal process. Several studies show that these batteries have energy levels and that they can be reused in other applications outside of the car as a power supply. This option of using recovered batteries is known as the second life of the battery (SLB). The reuse of batteries generates options to supply power on a large scale and with this reduce the pollution that these batteries can generate, especially in our country that does not have an optimal recycling process. This chapter presents the design of a methodology for the implementation of second life in Ecuador considering the use of NiHm batteries in HEV. For the design of the methodology, two possible scenarios for its implementation are analyzed. Scenario 1 is the use of NiHm batteries to supply energy to laboratories of a University in the city of Cuenca and scenario 2 shows the use of NiHm batteries as an additional energy source at the Airport of Santa Cruz present in the Galapagos Islands.

**Keywords:** Second Life Battery (SLB), NiHm, Hybrid cars, Energy, Battery Pack

## 1. Introduction

In Ecuador, the presence of hybrid vehicles in recent years has had a growth in sales. These vehicles have a traction system based on an electric motor coupled to a combustion engine and a high voltage battery as an energy source. The presence of more than 20,000 vehicles sold from 2009 to 2020 (**Figure 1**) [1], raises concern especially about the high-voltage battery, which in some cases has already reached its useful life. In the country there are no recycling procedures and regulations for the treatment of the different chemical compounds of this type of batteries, this may generate a large environmental impact in the future.

These batteries are generally made of Nickel Hydride metal (NiHm), according to studies presented in [2–4] indicate that their reuse is possible to store energy in a SLB process. Its application in second-life energy storage systems would help in our country to generate NiHm battery reuse projects that in the future would reduce the environmental impact and pollution that these batteries could cause.

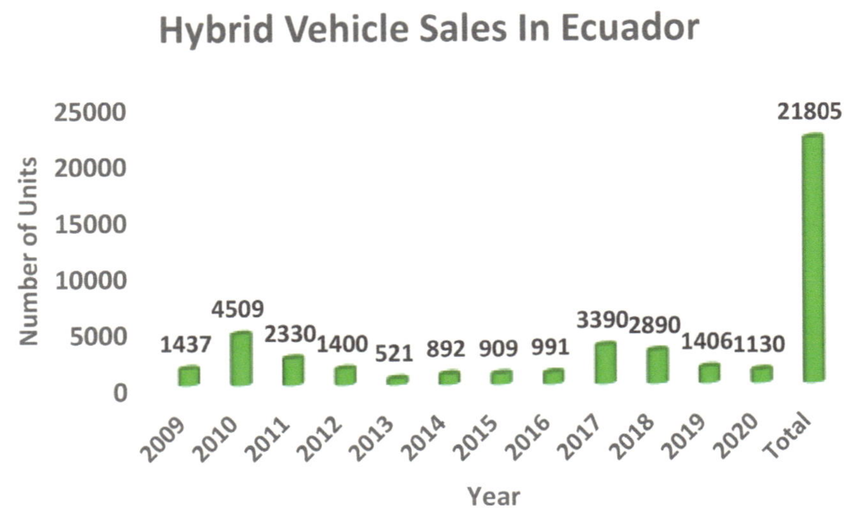


**Figure 1.**
*Sells of hybrids car in Ecuador.*

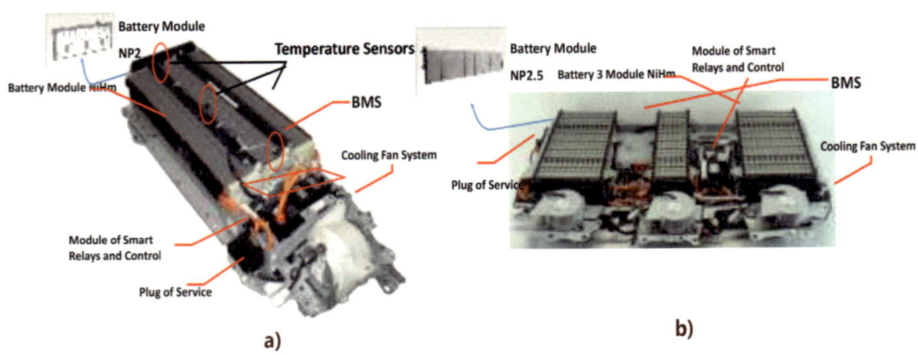


**Figure 2.**
*Types of NiHm batteries in HEV presents in Ecuador. a) Prius C battery, b) Toyota Highlander battery.*

The high-voltage batteries for hybrid vehicles with the most presence in Ecuador are of the NiHm type (**Figure 2**). The battery set is made up of several modules connected in series, these modules are of type NP2 and NP2.5. In addition, it has a control system based on high voltage relays, a cooling system that is activated based on the temperature sensors and a battery management module that is responsible for the supervision of each of the modules.

The NiHm batteries are robust systems that has higher capacity and low maintenance compared to nickel cadmium battery (NiCd). Several laboratories have developed feasibility studies on the use of NiHm batteries in their second life. The studies focus on analyzing the behavior of reused batteries vs new batteries of different chemical compounds, the results show that the use of these batteries in stationary systems is possible with good results [5]. If we consider that SBL systems with lithium-ion batteries are currently being implemented in different countries and show high efficiency and are less expensive, it is due to a high presence of Electric vehicles in these countries [6–15]. But if we compare them with our country, the greatest presence of batteries in the recycling process is that of hybrid vehicles based on NiHm, which allows us to generate an alternative for reuse in energy storage systems.

In continental Ecuador, the battery energy storage system is limited, there are no projects that are developed in this line. In the Galapagos Islands there are projects that consider energy storage based on lead-acid batteries and lithium ion batteries. These projects represent a little less than 5% of all generation capacity on the islands [16]. For all this and the presence of hybrid vehicles, it is proposed to use NiHm batteries as energy storage as an additional supply and with this contribute to caring for the environment and reduce the use of fossil fuels for electricity generation.

This chapter contributes to the search for new sources of energy power based on the use of batteries present in hybrid vehicles and that due to their operating condition can be adapted to a second life process, in the same way it is provided with a methodology for its implementation. Through its implementation, it seeks to improve the recycling processes of this type of batteries and contribute to the care of the environment.The structure of this chapter is as follows: in Section 2 a description of methodology proposed for the implementation of SLB based in NiHm battery of HEV is presented. The Section 3 presents two possible scenarios for the implementation of SBL with NiHm batteries where a comprehensive analysis is presented on the energy demand of each scenario and how much energy storage system with NiHm batteries can contribute if it were to be implemented in each case, considering the proposed methodology; while the conclusion is presented in Section 4.

## 2. Methodology proposed

Recycling processes in Ecuador are still in an early process. According to 2017 demographic statistics only four percent of the total garbage generated in Ecuador was recycled. This was the last year statistics about waste disposal were published. It's not clear how much of this four percent of recycled material has a second life process but most of it it's just simply classified, turned into pallets and sent recycling facilities. In the case of steel, paper and glass, these raw materials are processed in local companies and for most other material such as plastic, aluminum and electronics are sent to recycling facilities in countries overseas.

Such is the case of hybrid modules extracted from hybrid vehicles. Now a days a battery is not a whole individual pack as we seen before with Lead acid batteries but rather a composition or various cells. These cells can actually be replaced in case they are damaged prematurely and will give the whole battery pack a new life. This change will allow the rest of the cells to complete their life span as they may not have been affected. This practice is not sponsored by the manufacturer for safety and commercial purposes, but in this section, we will discuss a methodology proposed in Ecuador to properly and safely do a restoration for the battery de NiHm presents in HEV in order to prolong its life and generate a second life system for using as additional power supply.

The proposed methodology consists of 5 stages (**Figure 3**). The first stage involves the removal of the defective NiHm battery pack from the vehicles, this involves searching for cars with these problems and battery packs that are in recycling shops or automotive dealers. The second stage consists in disarming the batteries to check their shape and discard internal failure or explosion of the cell. In the third stage, the batteries are tested to check their health and charge state, this allows establishing a battery classification and verifying whether or not it is useful in a second life process. Stage four, generates processes for the assembly of battery packs with reestablished cells and the reconfiguration of different elements such as the BMS and the recharging system. Finally, stage 5 shows the implementation of battery packs that can be used as an additional power source in different applications.

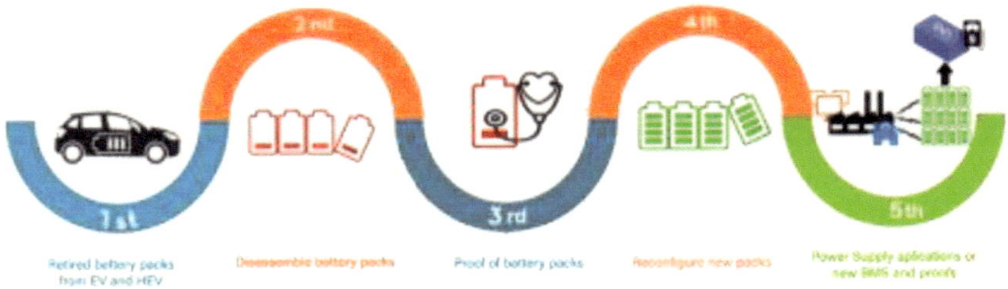


**Figure 3.**
*Methodology proposed for implementation of SLB in Ecuador.*

**Figure 4.**
*Methods for obtain and retired battery packs NiHm.*

## 2.1 Retired battery packs

In this stage we must consider several aspects; the consumer is the first link to building a methodology towards car batteries second life in Ecuador. As the number of HEV grow, costumers of such goods have to be aware of the life span of the battery an of how they should be disposed of after they have completed their life cycle in the vehicle. Costumer should understand that under any circumstance they are not allowed to take their battery home unless they have a special permission to do it. To reinforce this practice, among costumers, applied should be applied policies of change of battery generated by the brand with a presence in the country and laws by part of government that allow good process of recycling for these components. In order to obtain batteries, it is necessary to link with local companies that offer battery replacement and with the dealers of the different brands, in addition to the recycling warehouses, which have these batteries but do not know their origin. All these actions are presented in the **Figure 4**.

## 2.2 Disassemble battery packs

The stage involves the battery disassembly process (**Figure 5**), considering different safety standards, procedures and steps are generated to follow for disassembling the different types of batteries and recycling of parts.

## 2.3 Proof of battery packs

This stage consists of carrying out different charging and discharging processes to select the batteries that are in good condition and carry out a process of cell

**Figure 5.**
*Battery disassembly process.*

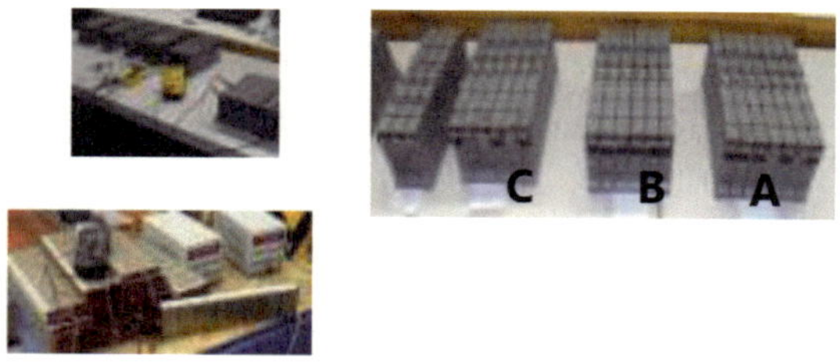


**Figure 6.**
*Process of selection and battery test.*

classification (**Figure 6**). These processes will allow the cells to be classified into three categories, A good condition, B intermediate condition, C damaged cells.

## 2.4 Reconfigure new packs

In this stage, new battery packs are built with the recovered cells and the BMS system is reconfigured in the same way, the battery charging method is selected and power ranges to work are analyzed (**Figure 7**).

## 2.5 Power supply application

This stage implies the use in places and scenarios where they can be installed. Each battery pack is important because it contributes a percentage amount of power at different times of the day such as maximum consumption peaks at certain hours. By restoring the damaged battery cells the conventional network power consumption can be reduced (**Figure 8**).

**Figure 7.**
*Reconfigured new battery packs.*

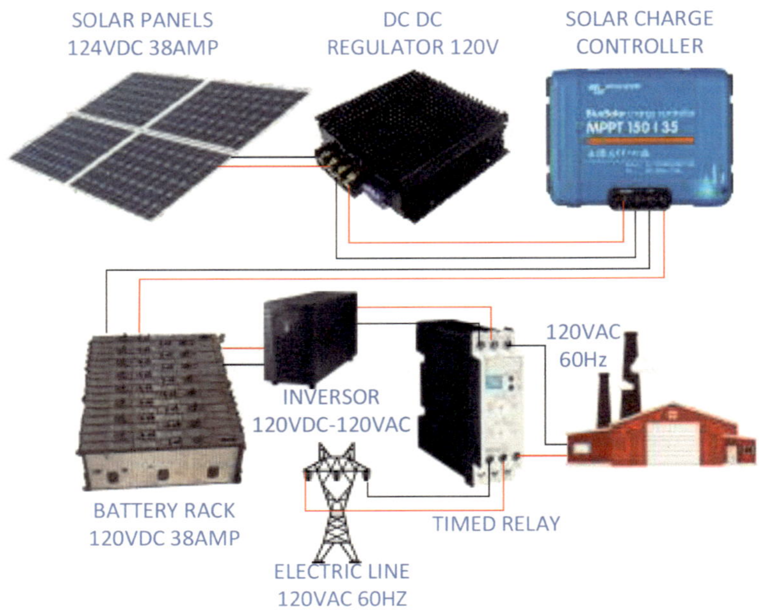


**Figure 8.**
*Laboratory power configuration with battery packs.*

## 3. Implementation of SLB based in NiHm battery analysis of two scenarios

In Ecuador the electricity sector is considered strategic due to its direct relationship with productive development. In recent years there have been changes in the generation of electricity according to the indicators shown in [17], the sources of renewable energy generation have increased, which is currently well seen by various productive sectors of the country.

The energy requirements at the country level are supplied by hydraulic energy followed by thermal energy and with some generation stations by turbo steam and gas, with solar and wind energy being the ones with the lowest contribution in power generation (**Figure 9**).

The Electricity Regulation and Control Agency -ARCONEL-, according to article 15, of the Organic Law of the Public Electricity Service, has the competence to carry out technical studies and analysis for the elaboration of electricity rates in the public service. The prices and values that to date that are charged in Ecuador

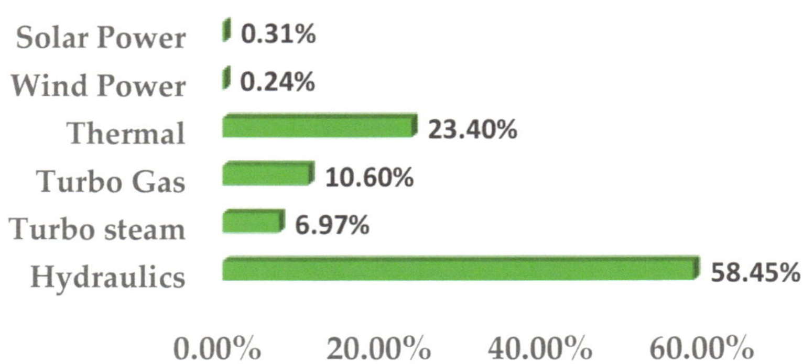


**Figure 9.**
*Forms of power generation in Ecuador.*

are categorized: the first category is the residential sector with energy consumption that exceeds 500 kilowatt hours (kWh) per month, which will pay a flat rate (fixed) of 10 cents per kWh. The second group is the dignity rate for consumptions that do not exceed 500 kWh will pay 4 cents per kWh and the third group is the industrial sector with a rate of 60 cents per kWh this if we speak on a continental level.

The importance of generating alternative options for energy storage systems based SLB recycled batteries can be beneficial to the sustainable development of the country and generate interest to circular economy of systems of batteries for large scale implementation.

This section presents a study of an implementation of SLB in two scenarios using NiHm batteries as energy storage. For the reconditioning and use of the batteries, the methodology presented in the previous section was used. The goal is to be able to analyze its field of application and its energy contribution to reduce the consumption of the conventional network, this would generate a reduction in the payment of electricity value and less energy would be spent from the main power network. The first scenario presents the study of implementation of a system SLB in Galapagos Island. The Galapagos Islands, is a province of Ecuador, they are located 972 kilometers west of the Ecuadorian coast. The capital is Puerto Baquerizo Moreno and it depends directly on the National Government. It is made up of 13 large volcanic islands, 6 smaller islands and 107 rocks and islets (**Figure 10**).

In the Islands since 2007, the governments and the United Nations Development Program (UNDP have promoted the project called "zero fossil fuels", in previous years the main source of generation was thermal, which generated excessive spending on diesel fuel and pollution in the area Currently, different projects based on renewable energy have been built with very promising results [18]. The main sources of generation in the islands are thermal, wind, solar and hybrid.

Energy demand is too high at certain hours, generating peaks that the current energy supplies are not able to handle, therefore the use of additional energy supply systems can be a help to be able to solve these inconveniences.

In the Galapagos Island there are different public and private institutions the **Figure 11** presents differents publics institutions and your monthly electricity consumption before the pandemic and the amount to pay.

The Isla San Cristóbal Airport in Galapagos is the second main passenger terminal in the Archipelago and has an electrical energy consumption of 14,000 kWh approximately per month and represents an expense of $ 1,200 per month

**Figure 10.**
*Galapagos Island geographic map.*

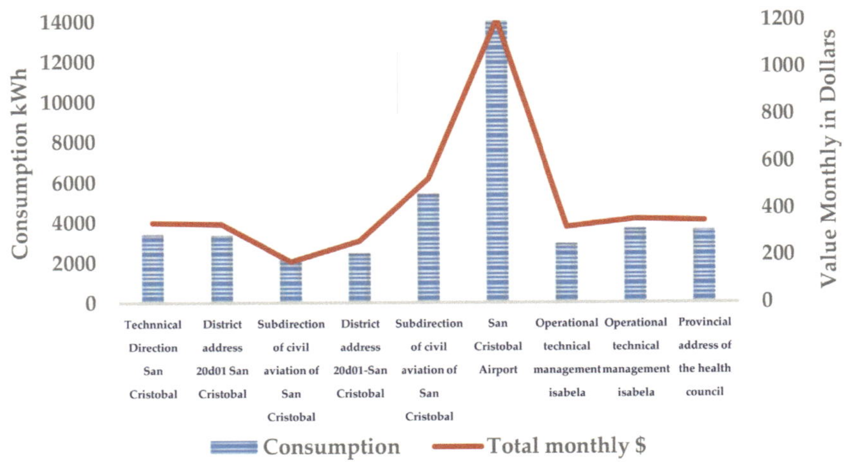


**Figure 11.**
*Energy consumption of public entities in the Galapagos Islands by approximate monthly.*

this data is before the pandemic. Considering the methodology proposed in the previous section, 5 battery racks are implemented and each one with 120 volts dc and with a current of 30 amperes, the design conditions for the system would be as presented below.

Each rack provides a power of 3.6 kW, the SLB system would have a power of 18 kW. To be able to generate a power in this range, the design of a photovoltaic system is needed that will serve to generate the amount of power necessary to charge the battery racks. The photovoltaic system is made up of 50 solar panels of 370 watts a BSP370M system is used, which is a model available in our country. The system is shown in **Figure 12**.

The photovoltaic system proposed generates 18500 Watts. In the Galapagos Islands there is sun throughout the year with a duration of more than 10 hours a

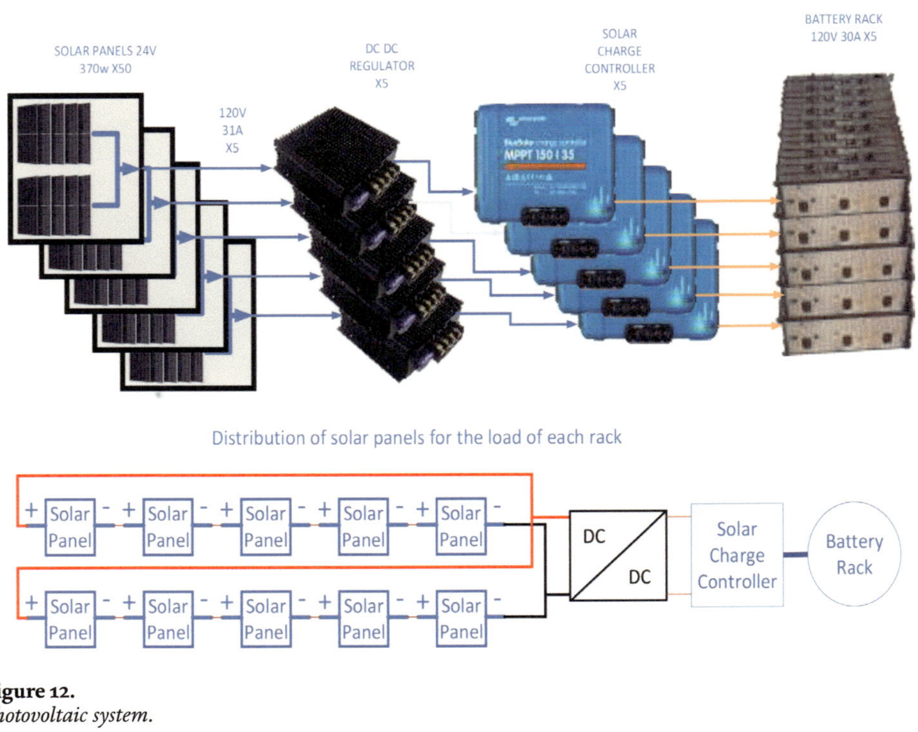


**Figure 12.**
*Photovoltaic system.*

| Battery Racks | Voltage and Current | Power of Panels (w) | N. of Panels | Hours per month | Amount kWh |
|---|---|---|---|---|---|
| 5 | 120 V -30 A | 370 | 50 | 240 | 4440 |

**Table 1.**
*Values and contribution of SBL.*

day of exposure, which allows an optimal generation by the system. With all these data, **Table 1** presents an analysis of this scenario and the contribution that the SLB system has with respect to its possible implementation.

If we compare the results obtained with the system SLB and the actual consumption of the airport, we can analyze that the contribution of system represents the 31.71% of the total of kWh by month. This would allow us to free up a demand of almost 4440 kWh that can be used by other users within the islands (**Figure 13**).

The cost for implementation is presented in **Table 2**. This section does not consider the cost and design of the DC-AC inverters to generate alternating current.

This proposal would cover its implementation expenses in 36 months, considering that the proposed BSL system contributes 4440 kWh at a differentiated rate of 0.086 US cents per kWh (according to the payment rate within the islands) and would generate a saving of US $ 4582.08 per 12 months. Undoubtedly it generates interest for its future implementation within this analyzed scenario.

The selection of the panels for this project was carried out with load and unload tests to verify the state of charge (SOC) of each module. With the panels in better

**Figure 13.**
*Analysis of contribution of SLB system first scenario.*

| Devices | Cost Unit | Total |
|---|---|---|
| 50 solar panels BSP370M | $ 204.50 | $ 10200 |
| 80 Battery Cells NiHm | $3 by recycling | $ 240 |
| 5 Solar charger controllers | $ 61 | $ 305 |
| 5 DC-DC regulators | $35 | $175 |
| Total | | $ 10920 |

**Table 2.**
*Cost of implementation.*

condition, we make up the racks used for the supply of energy, but in the future where it will be carried out with a higher number of modules in the new racks, the selection will be made by means of an electrochemical model.

For this model, certain values are considered that establish the aging of the batteries: Temperature (T), Depth of discharge (DOD), state of charge (SOC) and current rate (I). In Eq. (1) the dependence of the factors on the loss capacity is described [19].

$$C_{loss} = f(I, V, DOD, T, t) \tag{1}$$

The fading of the storage capacity of each cell is a function of current (C) and DOD factors. To obtain these factors, charge and discharge tests should be performed in percentages of a full discharge of C, 50% SOC, and 100% DOD. In Eqs. (2)–(5) the degradation is adjusted in relation to the aging factors [19].

$$I_{ef} = \theta_1 \cdot I^2 + \theta_2 \cdot I + \theta_3 \tag{2}$$

$$V_{ef} = \theta_4 \cdot V + \theta_5 \tag{3}$$

$$DOD_{ef} = \frac{\log_{10}(DOD)}{2} \tag{4}$$

$$T_{ef} = \frac{e^{\frac{\theta_6}{T}}}{e^{\frac{\theta_6}{298}}} \tag{5}$$

The parameters $\theta_1, \theta_2, \theta_3, \theta_4, \theta_5$ and $\theta_6$ should be determined according to the type of battery.

Once you have the loss of capacity due to aging, the state of health (SOH) is expressed in Eq. (6), it is calculated between the initial capacity of the battery and the capacity obtained in the tests.

$$SOH = \frac{Cap}{Cap_{ini}} \tag{6}$$

The second scenario analyzes the implementation of a SLB system in the automotive mechanics laboratories of the Universidad del Azuay located in the city of Cuenca, in the south of the country. The laboratory has a monthly consumption of 3600kWh each kWh has a value of 0.095 cents of a dollar that per year means an expense of $ 4104 dollars for the University. It is considered to place a 120 V battery rack at 35 amps. The system supplies a power of 4.2 kW. For charging the battery pack using solar panels can be used panels from 24 V to 455 Watts for our system used models JAM72S20–455 MR and consider the structure of the **Figure 14**.

It takes 10 solar panels, two groups in parallel with five panels in series this supplies the voltage and current requirement for battery charging. The arrangement of the panels in series with their reference values and regulators is shown in

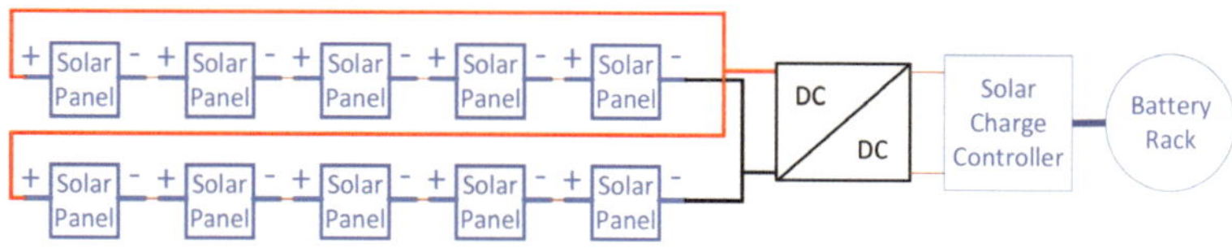


**Figure 14.**
*Charging structure with solar panels.*

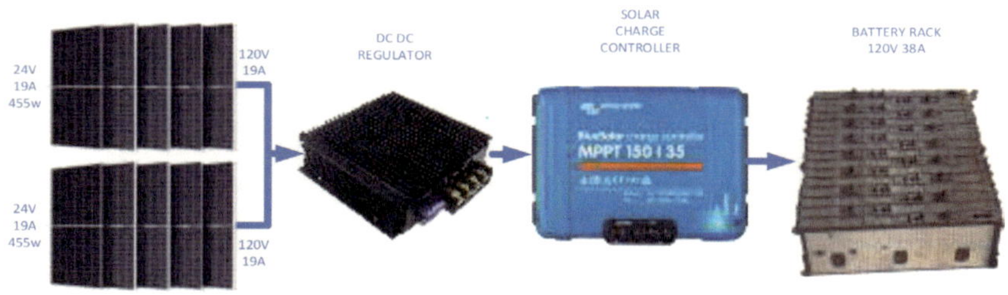


**Figure 15.**
*Components of the solar powered charging unit.*

**Figure 15**. The voltage of solar panels varies depending on factors such as ambient temperature, which is why regulators are used that take care of the useful life of the batteries at the time of charging.

Cuenca is a city that is located in the Ecuadorian highlands, it usually has 12 hours of sun but the climatic situations are different, so we consider 8 hours a day in which the solar panel system can generate. The photovoltaic system proposed generates 4500 Watts. The contribution in kWh of this system SLB in second scenario represent 1092 kWh and is shown in the **Table 3**.

The implementation of this system contributes 30% of the consumption in kWh, which represents a good contribution to reduce consumption costs and that the 1092 kWh that this represents can be used by other clients of the electricity company (**Figure 16**).

In this case, the implementation costs would be covered after 26 months. Considering that the proposed BSL system contributes 1092 kWh at a differentiated rate of 0.095 US cents per kWh (according to the payment in the country) and would generate a saving in 1244.88 by 12 months.

| Devices | Cost Unit | Total |
|---|---|---|
| 10 solar panels JAM72S20–455 MR | $ 250 | $ 2500 |
| 16 Battery Cells NiHm | $3 by recycling | $ 48 |
| 1 Solar charger controllers | $ 61 | $ 61 |
| 1 DC-DC regulators | $35 | $35 |
| Total | | $ 2644 |

The following analysis describes the effects caused by the use of this type of batteries in second life considering the two scenarios studied. A comparison of the two scenarios with and without SBL system is made. For this analysis, the need for energy consumption of the scenario is considered, which is 14kWh of this amount, the contribution of the SLB system designed for this scenario is 4.4kWh at the energy cost level, which means a money saving of almost 20% with just 80 cells of NiHm batteries that generate a power of 18 kW (**Figure 17**).

| Battery Racks | Voltage and Current | Power of Panels (w) | N. of Panels | Hours per month | Amount kWh |
|---|---|---|---|---|---|
| 1 | 120 V -35 A | 455 | 10 | 240 | 1092 |

**Table 3.**
*Value of contribution of SLB system in second scenario.*

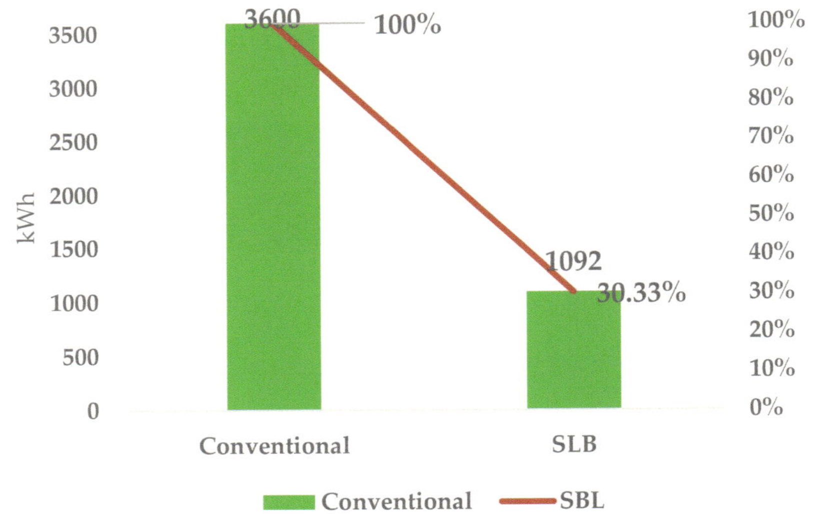


**Figure 16.**
*Analysis of contribution of SLB system second scenario.*

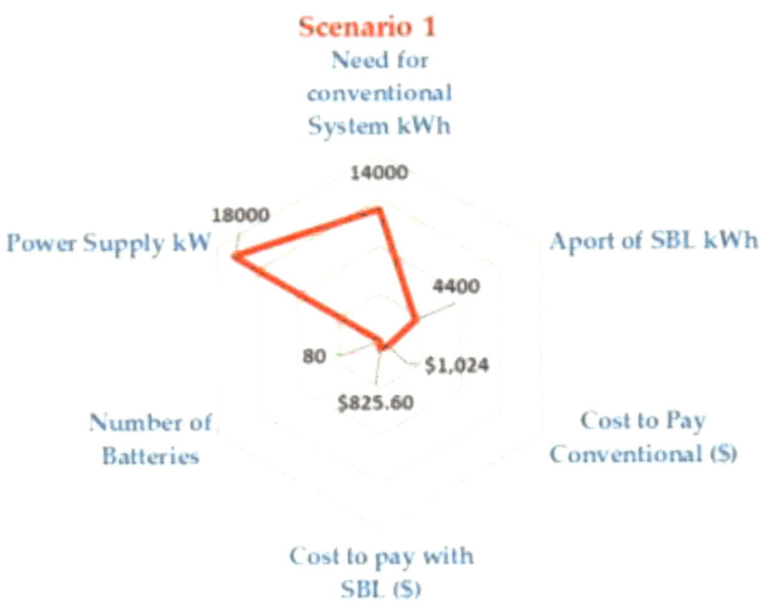


**Figure 17.**
*Analysis and comparison of first scenario with and without SBL system.*

For the second scenario, the need for energy consumption is 3.6 kWh and the contribution of the SLB system designed for this scenario is 2.5 kWh at the energy cost level, which means a money saving of almost 58% with just 16 battery cells. NiHm that generate a power of 4.2 kW (**Figure 18**).

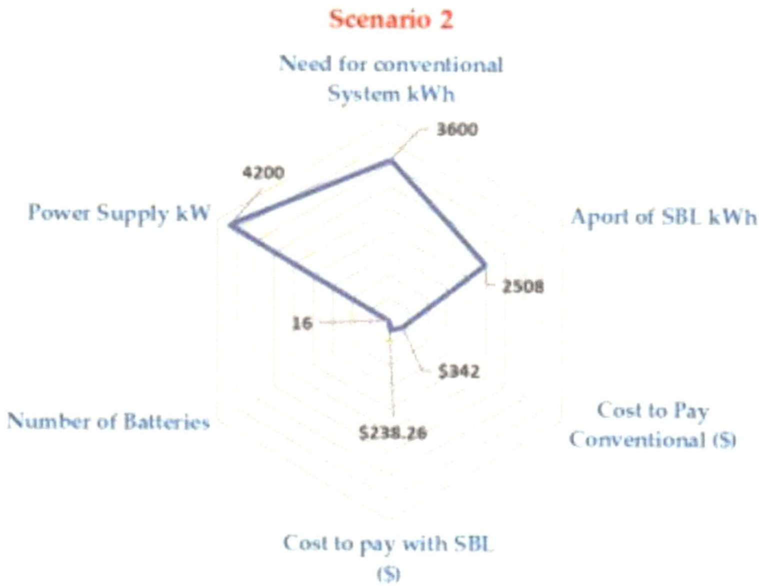


**Figure 18.**
*Analysis and comparison of the second scenario with and without SBL system.*

## 4. Conclusions

The introduction of hybrid vehicles in the country and especially the treatment of battery packs mean that this type of study allows the generation of new lines of research that allow the development of projects based on the second life of batteries. In our country very little is said about the advantages in the use of NiHm batteries present in these vehicles to be used as energy accumulators or as second life systems and the environmental impacts that this can generate.

This study focuses on generating methodologies that allow us to improve the quality of recycling these elements and generate a special treatment for their recovery. In addition, in contributing to society and being able to arouse the interest and attention of local and national government agencies and in conjunction with the academy to be able to generate circular economy projects and recycling regulations for this type of batteries and seek financing mechanisms for their implementation in strategic areas of the country.

The country is not a producer of this type of technology, but if we use correct recycling methods and develop optimal systems with second-life materials in the future, we can become a model to follow for the rest of Latin American countries, considering that hybrid vehicles continue to enter and increase. Sales and the transition to electric vehicles is getting closer.

## Acknowledgements

This project has been promoted by the Research Department of the University of Azuay in the 2020-0167 project and by the company Expertronics Ecuasolar.

## Conflict of interest

The authors declare no conflict of interest.

## Author details

Efrén Fernández Palomeque[1*], Diego Rojas Hiedra[1], Daniel Cordero[1] and Martín Espinoza[2]

1 ERGON Center Research, University of Azuay, Cuenca, Ecuador

2 Expertronics Ecuasolar, Quito, Ecuador

*Address all correspondence to: efernandez@uazuay.edu.ec

## References

[1] AEADE. Sector automotriz en cifras. 2021. Available from: https://www.aeade.net/boletin-sector-automotor-en-cifras/ [Accessed: 2021-01-21]

[2] C. C. Chan, "The State of the Art of Electric, Hybrid, and Fuel Cell Vehicles," in *Proceedings of the IEEE*, vol. 95, no. 4, pp. 704-718, April 2007, doi: 10.1109/JPROC.2007.892489.

[3] N. Pinsky, L. Gaillac, A. Mendoza, J. Argueta and T. Knipe, "Performance of advanced electric vehicle batteries in stationary applications," *24th Annual International Telecommunications Energy Conference*, 2002, pp. 366-372, doi: 10.1109/INTLEC.2002.1048682.

[4] E. Martinez-Laserna, I. Gandiaga, E. Sarasketa-Zabala, J. Badeda, D.-I. Stroe, M. Swierczynski, A. Goikoetxea,Battery second life: Hype, hope or reality? A critical review of the state of the art, Renewable and Sustainable Energy Reviews, Volume 93, 2018, Pages 701-718, ISSN 1364-0321, https://doi.org/10.1016/j.rser.2018.04.035.

[5] Zhao, Y.; Pohl, O.; Bhatt, A.I.; Collis, G.E.; Mahon, P.J.; Rüther, T.; Hollenkamp, A.F. A Review on Battery Market Trends, Second-Life Reuse, and Recycling. *Sustain. Chem.* 2021, *2*, 167-205. https://doi.org/10.3390/suschem2010011.

[6] Lluc Canals Casals, B. Amante García, Camille Canal, Second life batteries lifespan: Rest of useful life and environmental analysis, Journal of Environmental Management, Volume 232, 2019, Pages 354-363, ISSN 0301-4797, https://doi.org/10.1016/j.jenvman.2018.11.046.

[7] N. Mukherjee and D. Strickland, "Analysis and Comparative Study of Different Converter Modes in Modular Second-Life Hybrid Battery Energy Storage Systems," in *IEEE Journal of Emerging and Selected Topics in Power Electronics*, vol. 4, no. 2, pp. 547-563, June 2016, doi: 10.1109/JESTPE.2015.2460334.

[8] E. Hossain, D. Murtaugh, J. Mody, H. M. R. Faruque, M. S. Haque Sunny and N. Mohammad, "A Comprehensive Review on Second-Life Batteries: Current State, Manufacturing Considerations, Applications, Impacts, Barriers & Potential Solutions, Business Strategies, and Policies," in *IEEE Access*, vol. 7, pp. 73215-73252, 2019, doi: 10.1109/ACCESS.2019.2917859.

[9] E. Martinez-Laserna *et al.*, "Evaluation of lithium-ion battery second life performance and degradation," *2016 IEEE Energy Conversion Congress and Exposition (ECCE)*, 2016, pp. 1-7, doi: 10.1109/ECCE.2016.7855090.

[10] D. Strickland, L. Chittock, D. A. Stone, M. P. Foster and B. Price, "Estimation of Transportation Battery Second Life for Use in Electricity Grid Systems," in *IEEE Transactions on Sustainable Energy*, vol. 5, no. 3, pp. 795-803, July 2014, doi: 10.1109/TSTE.2014.2303572.

[11] A. Saez-de-Ibarra, E. Martinez-Laserna, C. Koch-Ciobotaru, P. Rodriguez, D. Stroe and M. Swierczynski, "Second life battery energy storage system for residential demand response service," *2015 IEEE International Conference on Industrial Technology (ICIT)*, 2015, pp. 2941-2948, doi: 10.1109/ICIT.2015.7125532.

[12] Faessler, B.; Kepplinger, P.; Petrasch, J. Field Testing of Repurposed Electric Vehicle Batteries for Price-Driven Grid Balancing. J. Energy Storage 2019, 21, 40-47.

[13] Faessler B. Stationary, Second Use Battery Energy Storage Systems and

Their Applications: A Research Review. Energies. 2021; 14(8):2335. https://doi.org/10.3390/en14082335

[14] Gohla-Neudecker, B.; Maiyappan, V.S.; Juraschek, S.; Mohr, S. Battery 2nd Life: Presenting a Benchmark Stationary Storage System as Enabler for the Global Energy Transition. In Proceedings of the 2017 6th International Conference on Clean Electrical Power (ICCEP), Santa Margherita Ligure, Italy, 27-29 June 2017; pp. 103-109.

[15] Viswanathan, V. V., Kintner-Meyer, M., 2011. Second use of transportation batteries: Maximizing the value of batteries for transportation and grid services. IEEE Trans. Veh. Technol. 60, 2963-2970. doi:10.1109/TVT.2011.2160378

[16] Plan Nacional de Desarrollo toda una Vida 2018-2021 Cap 4 Expansión de la Generación https://www.recursosyenergia.gob.ec/wp-content/uploads/2020/01/4.-EXPANSION-DE-LA-GENERACION.pdf

[17] Agencia de control de recursos y energía. Panorama Eléctrico, datos junio 2020 Quito – Ecuador, noviembre 2020. Available from: https://www.controlrecursosyenergia.gob.ec/wp-content/uploads/downloads/2020/11/Revista-01-V2-compressed.pdf. [Accessed: 2021-04-21]

[18] Inter-American Development Bank (IADB); Ministry of Electricity and Renewable Energy (MEER). National Plan for Energy E_ciency; IADB; MEER: Quito, Ecuador, 2017; Available online: https://www.cnelep.gob.ec/ plan-nacional-eficiencia-energetica/ (accessed on 7 August 2019).

[19] Canals Casals L., Amante García B., González Benítez M.M. (2017) Aging Model for Re-used Electric Vehicle Batteries in Second Life Stationary Applications. In: Ayuso Muñoz J.L., Yagüe Blanco J.L., Capuz-Rizo S.F. (eds) Project Management and Engineering Research. Lecture Notes in Management and Industrial Engineering. Springer, Cham. https://doi.org/10.1007/978-3-319-51859-6_10

Chapter 4

# Current Status and Prospects of Solid-State Batteries as the Future of Energy Storage

*Marm Dixit, Nitin Muralidharan, Anand Parejiya, Ruhul Amin, Rachid Essehli and Ilias Belharouak*

## Abstract

Solid-state battery (SSB) is the new avenue for achieving safe and high energy density energy storage in both conventional but also niche applications. Such batteries employ a solid electrolyte unlike the modern-day liquid electrolyte-based lithium-ion batteries and thus facilitate the use of high-capacity lithium metal anodes thereby achieving high energy densities. Despite this promise, practical realization and commercial adoption of solid-state batteries remain a challenge due to the underlying material and cell level issues that needs to be overcome. This chapter thus covers the specific challenges, design principles and performance improvement strategies pertaining to the cathode, solid electrolyte and anode used in solid state batteries. Perspectives and outlook on specific applications that can benefit from the successful implementation of solid-state battery systems are also discussed. Overall, this chapter highlights the potential of solid-state batteries for successful commercial deployment in next generation energy storage systems.

**Keywords:** solid electrolyte, composite cathode, lithium-ion, batteries, lithium anode

## 1. Introduction

The dawn of the 21$^{st}$ century coincided with the global civilization leapfrogging into the digital age. At the core of this digital revolution was the rapid adoption and wide deployment of lithium-ion batteries (LIBs) [1]. Ever since, both industries and the scientific community have been engaged in the quest for unlocking the secrets for developing batteries with more energy, -power, -life, and -safety. In the recent years, the battery R&D community is once again at the cusp of another technological revolution that could redefine the energy storage sector for decades to come. This revolution is largely fueled by the rapidly advancing efforts into a new battery technology called all Solid-State Batteries (ASSBs) [2]. While conventional lithium-ion batteries have enjoyed unprecedented levels of research and industrial attention directed towards every aspect of their constituents, SSBs have largely been on the backburner. For decades, the battery research sector has been constantly attempting to integrate high-capacity lithium metal anodes (3860 $mAh.g^{-1}$) to advance energy density targets of liquid electrolyte-based LIBs [3]. However, in such conventional batteries, the hydrocarbon derived organic liquid electrolytes pose significant safety and performance related challenges with lithium metal anodes and has also been a significant barrier preventing wide commercial deployment [4]. With renewables and electric vehicles (EV) set to dictate the timeline for the next industrial revolution, the battery R&D community has come to a profound consensus that conventional lithium-ion batteries are nearing their upper limits of performance [5]. The initial interest towards SSBs was based on the use of solid electrolytes (SE), as they are potentially thought to provide a straightforward approach towards realizing the safe integration of high-energy lithium metal anodes. In recent years, new classes of high-performance solid electrolytes have emerged with high room-temperature conductivities ($\sim$1 mS.cm$^{-1}$) [6] comparable to that of conventional liquid electrolytes and high lithium-ion transference numbers ($\sim$1 for inorganic SE) [7]. This coupled with novel processing approaches targeting interfacial tuning and optimizations are propelling the new generation of SSB systems towards garnering a widespread support for commercial adoption. With the EV demand projected to skyrocket in the next few years, the need for the next-generation high energy batteries that would power these advanced automotive platforms is also growing [1, 8–10]. To this extent, several large-cap automotive companies including Toyota, Volkswagen, General Motors, Hyundai, and Ford have already made major investments in SSB technology companies with the aim to achieve full commercial deployment in the first half of the 21$^{st}$ century.

Despite the optimism and promises around all SSBs, there are several major challenges pertaining to each cell component and processing approach that needs further optimizations to achieve the overarching performance goals. For SSBs, there can be several material-level issues that can cause major cell-level catastrophic failures. We had recently reported that an ideal solid-state battery (**Figure 1a**) that delivers a high energy density should consist of the following [11] – (i) a high-capacity thin lithium metal anode/seed layer (thickness $\sim$1-5 μm seed layer + 15-40 μm plated from the cathode), (ii) a stable solid electrolyte with high ionic conductivities (thickness $\sim$1-20 μm, ideally dry), (iii) a cathode composite with optimized loading and tailored architecture (thickness $\sim$45-200 μm) [12]. In order to achieve high energy densities in SSBs, it is generally understood that the cathode should be the most voluminous part. Current collectors employed in an SSB can generally be <10 μm thick applied in the form of thin coatings to mechanically robust electrodes [13]. Additionally, SSBs also offer the possibility of cell stacking in two different configurations depending upon performance requirements – conventional stacking and bipolar stacking [14]. Bipolar stacking in a single package using bipolar current collectors decreases the packing volume thus increasing the volumetric energy

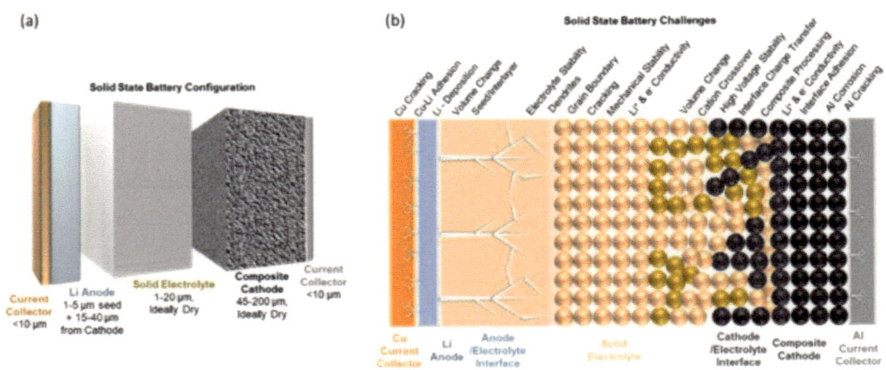


**Figure 1.**
*(a) An ideal solid-state battery with optimized configuration and (b) remaining material level challenges in solid state batteries.*

density. Though, the requirements to achieve high performance SSBs appear quite straight forward, significant fundamental issues and challenges pertaining to specific cell components remain unresolved in the SSB R&D space. **Figure 1b** depicts a schematic providing a brief overview of some underlying challenges that warrants systematic investigations and mitigation strategies to enable the wide commercial deployment of SSBs. From the aforementioned figure, it should be noted that interfacial properties play a dominant role in determining the final performance delivered by SSBs [15, 16]. For example, if we consider the simple interface between the anode and the copper current collector, some of the challenges encountered may include (i) current collector corrosion/cracking – due to the repeated mechanical cycling stresses and Li plating related side reactions, (ii) Cu-Li adhesion – improper adhesion between these dissimilar metals can occur due to oxide layer formation during processing or due to cyclic mechanical and electrochemical loads during charge/discharge, (iii) volume change – drastic volumetric expansion/contraction can occur during charge/discharge processes at the anode interfaces owing to the soft and pliable nature of Li metal, etc. Similarly, other interfaces between the lithium metal anode and solid electrolyte, solid electrolyte and cathode composite, cathode composite and positive current collector also suffer from interface related issues which can hamper final cell performance. Additionally, the component specific bulk property variations could also be crucial factors that affect SSB performance and thus demands systematic investigations leading to strategic solutions prior to commercial adoption.

In this chapter, we discuss the critical challenges, recent advances, and avenues for improvement for the various classes of cathodes, solid electrolytes and anodes that would facilitate the commercial adoption of next generation SSBs. We also discuss the key processing and fabrication criteria for assembling full SSBs cells with our recent modelling endeavors into achieving practical cell level energy densities. Finally, we discuss some perspectives on the challenges that remain unsolved as well as the future trends in SSB development. Through this chapter, we envision, the readers would get a comprehensive understanding of the recent trends, remaining challenges, and a clear perspective of the future prospects of all Solid-State Battery R&D.

## 2. Cathode materials for SSB

The cathode materials play a pivotal role in the energy density, power density, cycle life, and calendar life of a battery. Oxide-based lithium-ion intercalation materials are the cathode of today's choice and can mainly be segregated into five board classes based on their crystal structure.

i. Spinel structure ($Li[M]_2O_4$ with M = Mn, ($Mn_{1-y/2}Li_{y/2}$) or ($Mn_{3/4}Ni_{1/4}$)),

ii. Layered structure ($Li[M]O_2$ with M = Co, Ni, ($Ni_xCo_{1-x}$) or ($Ni_xMn_yCo_z$)),

iii. Olivine structure $Li[M']PO_4$ with M' = Fe, Mn, Ni, Co or ($Fe_yMn_{1-y}$),

iv. β II -$Li_2FeSiO_4$,

v. Tavorite-type $LiFeSO_4F$,

Apart from these, sulfur and some organic compounds are also used as cathode materials in batteries. Every cathode material and its crystal structure have inherent advantages and disadvantages from the standpoint of electrochemical energy storage. The basic concept of battery intercalation chemistry is very old [17]. The first intercalation reactions involving solid hosts (graphite) and guest molecules or ions (sulfate ions) were shown by Schauffautl in 1841. However, in the 1960s, the intercalation materials gained attention due to the altering of their electronic and optical properties through guest ion intercalation [18–20]. The transition-metal disulfides and oxides (such as $MS_2$ and $WO_3$) were first investigated for intercalation of $H^+$, $Li^+$, and $Na^+$ ions [17]. It was noted that the intercalation of these monovalent ions into the crystal structure of $WO_3$ altered the electronic structure and conductivity, resulting in the material changing from an insulator to metal depending upon the amount and types of monovalent cation intercalated. These intercalation reactions were also accompanied by structural changes with modifications to crystal chemistry [21].

## 2.1 Crystal structure cathode materials

The lattice atom and its coordination domain are the basic structural unit of a crystal. The lattice atoms are arranged periodically in specific combinations (e. g. space group) for the formation of crystals. It should be noted that, in general, the electronic structure and interaction for the bond formation energy in the structure unit ultimately determines the intrinsic chemical and physical properties of crystals [22]. For lithium-ion and sodium-ion batteries, the cathode materials can be formed by combinations of Li, Na, transition metal, and anion structure units. It should be further noted that the crystal structure and chemical composition of the cathode materials play an important role in the ionic and electronic transport properties and conduction mechanisms. The five different crystal structures of cathode materials are displayed in **Figure 2** [23]. The layered structure cathode exhibits two-dimensional ionic and electronic conductivity and diffusivity (**Figure 2a**).

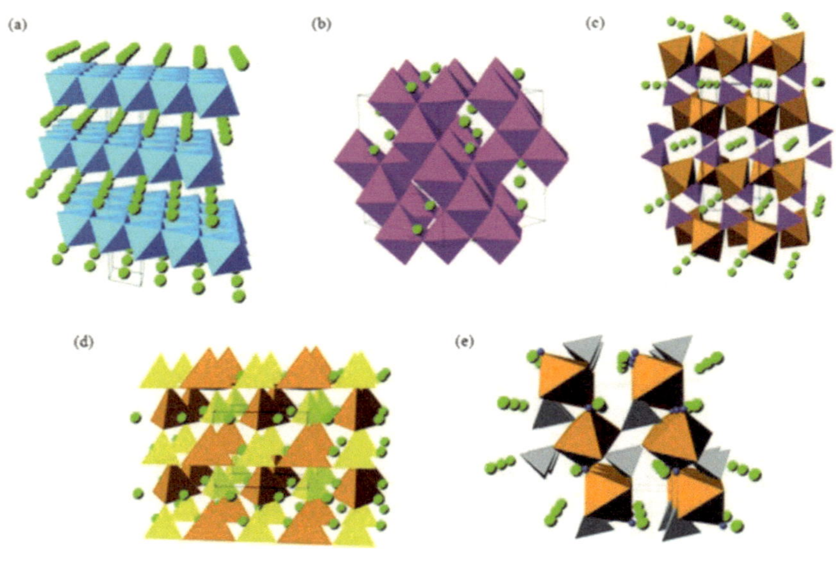


**Figure 2.**
*The crystal structures of major cathode materials: (a) layered α-$LiCoO_2$; (b) cubic $LiMn_2O_4$ spinel; (c) olivine-structured $LiFePO_4$; (d) β II -$Li_2FeSiO_4$; and (e) tavorite-type $LiFeSO_4F$. Li ions are shown as light green spheres, $CoO_6$ octahedra in blue, $MnO_6$ octahedra in mauve, Fe–O polyhedra in brown, $PO_4$ tetrahedra in purple, $SiO_4$ tetrahedra in yellow, $SO_4$ tetrahedra in gray, and in (e) fluoride ions in dark blue. Black lines demarcate one unit cell in each structure. (Reprinted with permission from Ref [23]. Copyright 2014, published by The Royal Society of Chemistry. This image is taken from the article titled "Lithium and sodium battery cathode materials: computational insights into voltage, diffusion and nanostructural properties", and it is under the Creative Commons Attribution 3.0 International License. To view a copy of this license, visit https://creativecommons.org/licenses/by/3.0/.).*

On the other hand, materials belonging to the spinel structure demonstrate three-dimensional ionic and electronic conductivity and ionic diffusivity (**Figure 2b**). In contrast, the olivine structure cathodes showcase preferably one-dimensional ionic conductivity and diffusivity and two-dimensional electronic conductivity (**Figure 2c**). The β II -$Li_2FeSiO_4$ and tavorite-type $LiFeSO_4F$ structures (**Figure 2d** and **e**) are not vigorously used in the lithium-ion battery cathode. The layer and spinel structure cathodes exhibit an interstitial type of mechanism of ionic conductivity and diffusivity. The olivine structure materials display vacancy migration type ionic conductivity and diffusivity. Unlike conventional lithium-ion battery (with liquid electrolyte), all solid-state battery (SSB) is impacted by dimensional conductivity, material hardness and mechanical properties which will be discussed in detail in the later sections.

### 2.2 Electronic and ionic transport properties

In a solid active cathode particle, the ion and electron moving together when charging or discharging the battery and this phenomenon is called ambipolar diffusion. Therefore, optimum ionic and electronic transport properties are prerequisites for high performances batteries, particularly SSBs. Nonetheless, it is very rare to get such good combinations in the presently available list of cathode materials. Some of the reported electronic and ionic conductivity and ionic diffusivity of spinel, layer, and olivine structure materials are compared and displayed in **Figure 3** and **Table 1** as a function of inverse temperature and lithium concentration, respectively.

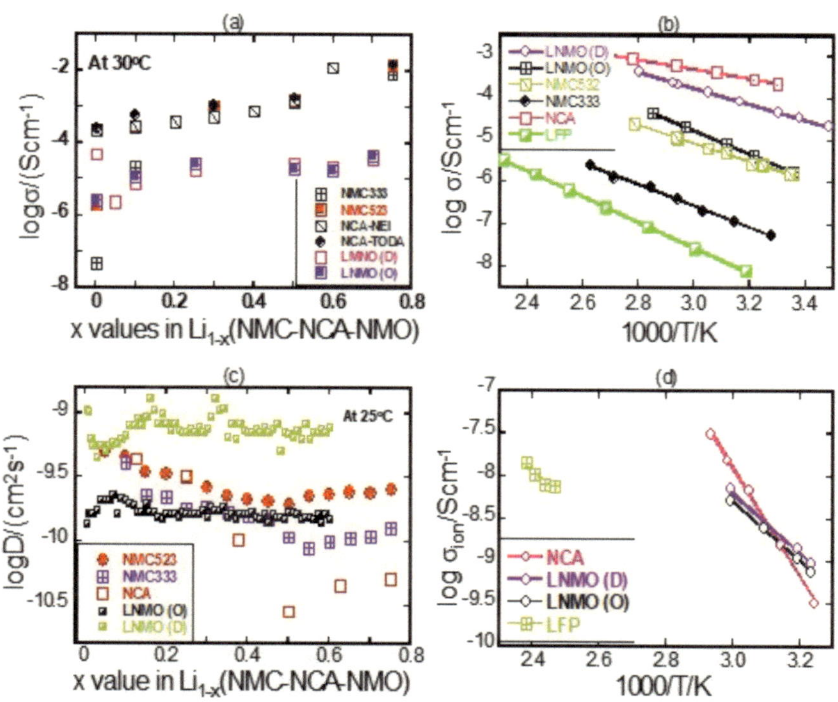


**Figure 3.**
*(a) Electronic conductivity of layer and spinel structure cathodes as a function lithium content, (b) Electronic conductivity of layer, spinel, and olivine structure cathode materials as a function of inverse temperature, (c) ionic diffusivity of layer and spinel structure cathode as a function of lithium content and (d) ionic conductivity of layer, spinel, and olivine structure cathode materials as a function of inverse temperature.* $LiFePO_4$ = *LFP,* $LiNixMnyCozO_2$ = *NMC,* $LiNi_{0.75}Co_{0.25}Al_{0.05}O_2$ = *NCA,* $LiMn_{1.5}Ni_{0.5}O_4$ = *LMNO (O), and* $LiMn_{1.5}Ni_{0.5}O_{4-\delta}$ = *LMNO (D). Each of the figures, a, b c and d were made by authors from the following references. [24] = LFP, [25] = (NMC333, NMC523), [26] = (NCA-NEI, NCA-TODA), [27] = (LNMO (O), LNMO (D)), [28] = (LNMO (O), LNMO (D)).*

| Technique Used | T/°C | Composition | Diffusivity ($cm^2/s$) | Reference | Structure |
|---|---|---|---|---|---|
| AC | 60 | $LiNi_{0.5}Mn_{0.2}Co_{0.3}O_2$ ($NMC_{523}$) | $\sim 3\times10^{-8}$ | [25] | |
| AC | 61 | $LiNi_{0.33}Mn_{0.33}Co_{0.33}O_2$ ($NMC_{333}$) | $\sim 5\times10^{-8}$ | [25] | |
| DC | 50 | $LiNi_{0.33}Mn_{0.33}Co_{0.33}O_2$ ($NMC_{333}$) | $\sim 7\times10^{-9}$ | [25] | |
| Depolarization | 25 | $Li_{0.9}Ni_{0.33}Mn_{0.33}Co_{0.33}O_2$ ($NMC_{333}$) | $\sim 4\times10^{-10}$ | [25] | |
| Depolarization | 25 | $Li_{0.25}Ni_{0.33}Mn_{0.33}Co_{0.33}O_2$ ($NMC_{333}$) | $\sim 10^{-10}$ | [25] | Layer |
| Depolarization | 25 | $Li_{0.9}Ni_{0.5}Mn_{0.2}Co_{0.3}O_2$ ($NMC_{523}$) | $\sim 5\times10^{-10}$ | [25] | |
| Depolarization | 25 | $Li_{0.25}Ni_{0.5}Mn_{0.2}Co_{0.3}O_2$ ($NMC_{523}$) | $\sim 2\times10^{-10}$ | [25] | |
| GITT | 25 | $LiNi_{0.33}Mn_{0.33}Co_{0.33}O_2$ ($NMC_{333}$) | $\sim 10^{-11}$ | [29] | |
| GITT | 25 | $Li_{0.25}Ni_{0.33}Mn_{0.33}Co_{0.33}O_2$ ($NMC_{333}$) | $\sim 10^{-10}$ | [29] | |
| GITT | 25 | $LiNi_{0.33}Mn_{0.33}Co_{0.33}O_2$ ($NMC_{333}$) | $\sim 10^{-9}$ | [30] | |
| GITT | 25 | $Li_{0.25}Ni_{0.33}Mn_{0.33}Co_{0.33}O_2$ ($NMC_{333}$) | $\sim 5x10^{-10}$ | [30] | |
| CV | 25 | $LiNi_{0.33}Mn_{0.33}Co_{0.33}O_2$ ($NMC_{333}$) | $\sim 5x10^{-14}$ | [31] | |
| CV | 25 | $LiNi_{0.33}Mn_{0.33}Co_{0.33}O_2$ ($NMC_{333}$) | $\sim 3x10^{-10}$ | [32] | |
| AC | 25 | $LiNi_{0.8}Co_{0.15}Al_{0.05}O_2$ (NCA) | $\sim 3\times10^{-10}$ | [26] | |
| DC Polarization | 25 | $LiNi_{0.8}Co_{0.15}Al_{0.05}O_2$ (NCA) | $\sim 2\times10^{-10}$ | [26] | |
| GITT | 25 | $Li_{0.875}Ni_{0.8}Co_{0.15}Al_{0.05}O_2$ (NCA) | $\sim 4\times10^{-10}$ | [26] | |
| GITT | 25 | $Li_{0.9}Ni_{0.5}Co_{0.5}O_2$ (NC) | $\sim 10^{-10}$ | [33] | |
| GITT | 25 | $Li_{0.9}Ni_{0.8}Co_{0.2}O_2$ (NC) | $\sim 10^{-8}$ | [34] | |
| First principle | 25 | $LiMn_{1.5}Ni_{0.5}O_{4-\delta}$ (LMNO (O)) | $\sim 10^{-9}$ | [35] | |
| DC | 50 | $LiNMnO_{4-\delta}$ (LMNO (D)) | $\sim 8\times10^{-9}$ | [27, 28] | Spinel |
| AC | 25 | $LiMn_{1.5}Ni_{0.5}O_{4-\delta}$ (LMNO (D)) | $\sim 5.1\times10^{-10}$ | [27, 28] | |
| GITT | 25 | $LiMn_{1.5}Ni_{0.5}O_{4-\delta}$ (LMNO (O)) | $\sim 5\times10^{-10}$ | [36] | |
| PITT | 25 | $LiMn_{1.5}Ni_{0.5}O_{4-\delta}$ (LMNO (O)) | $5\times10^{-12}$ -9 $\times 10^{-10}$ | [37] | |
| EIS | 25 | $LiNMnO_{4-\delta}$ (LMNO (D)) | $2.97\times10^{-15}$ | [38] | |
| PITT, EIS | 25 | $LiMn_{1.5}Ni_{0.5}O_{4-\delta}$ (LMNO (O)) | $10^{-12}$-$10^{-10}$ | [39] | |
| DC | 147 | $LiFEPO_4$ | $1.6\times10-9$ | [24] | Olivine |

**Table 1.**
*Comparison of ionic diffusivity of the selected major classes of cathode materials at specific temperature.*

It is seen from **Figure 3a** that the electronic conductivity of spinel (LMNO) and layer structure (NMC and NCA) cathodes show the discrete pattern of conductivity as a function of lithium content. It is discernible from **Figure 3b** that the layer structure materials exhibit a gradual increase of conductivity with increasing degree of delithiation. On the other hand, the electronic conductivity of the spinel structure can be manipulated by varying the degree of disorder and degree of delithiation. It should be noted that the spinel phase exhibits two crystallographic polymorphs, ordered and disordered depending on the distribution of Ni and Mn in the crystal structure. The ordered $LiMn_{1.5}Ni_{0.5}O_4$ (LMNO (O)) exhibit approximately fifteen times lower electronic conductivity than the disordered $LiMn_{1.5}Ni_{0.5}O_{4-\delta}$ (LMNO (D)) phase (**Figure 3a**) in the lithiated states. Also, the electronic conductivities of the ordered spinel LNMO (O), measured at a given temperature, increases gradually with increasing the degree of delithiation (**Figure 3a**). However, when x = 0.3 and beyond, the conductivity is almost leveled where it hardly changes with the removal of lithium in the measured range. It appears that partial lithium off-stoichiometric phases are favorable for better high-rate performances from comparing electronic conductivities for layered and ordered spinel cathodes. In disparity, the electronic conductivity of the disordered spinel (LNMO (D)) is reduced suddenly upon slight delithiation and falls to the level of the electronic conductivity demonstrated by lithiated ordered spinel phase.

Following which, the electronic conductivity exhibits alike trends for both the ordered and disordered phases (**Figure 3a**). The electronic conductivity of the olivine phase as a function of lithium concentration is not available in the literature. However, the electronic conductivity of olivine phase $LiFePO_4$ exhibits the lowest conductivity as a function of temperature (**Figure 3b**) and layer structure NCA shows the highest conductivity. The ionic conductivity and diffusivity of spinel phases appear to be favorable from the SSB standpoint compared to layered and olivine structure cathodes (cf. **Figure 3c** and **d** and **Table 1**).

### 2.3 Electrochemical performances and particle morphology

The energy and power density are important metrics to determine the scope of a specific application for cathode materials [35–39]. The theoretical and experimentally obtained capacity, operational average cell voltage (against carbon electrode), and energy density of major cathode materials are compared in **Table 2**. It is discernible from **Table 2** that all the phases of a particular crystal structure belonging to the same material family do not deliver the same energy density. The layer structure cathodes exhibit higher capacity compared to the spinel and olivine.

| Crystal structure | Composition | Capacity/mAh/g (theoretical/ experimental) | Average cell voltage (V) | Experimental energy density (Wh/kg) | References |
|---|---|---|---|---|---|
| Layered | $LiTiS_2$ | 225/210 | 1.9 | 399 | [40] |
| | $LiCoO_2$ | 274/148 | 3.8 | 562 | [41] |
| | $LiNiO_2$ | 275/150 | 3.8 | 570 | [42] |
| | $LiMnO_2$ | 285/140 | 3.3 | 462 | [43] |
| | $LiNi_{0.33}Mn_{0.33}Co_{0.33}O_2$ | 280/160 | 3.7 | 592 | [44] |
| | $LiNi_{0.8}Co_{0.15}Al_{0.05}O_2$ | 279/199 | 3.7 | 736 | [45] |
| | $Li_2MnO_3$ | 458/180 | 3.8 | 684 | [46] |
| Spinel | $LiMn_2O_4$ | 148/120 | 4.0 | 480 | [47] |
| | $LiMn_{1.5}Ni_{0.5}O_4$ | 148/140 | 4.1 | 574 | [48] |
| | $LiCo_2O_4$ | 142/84 | 4.1 | 344 | [49] |
| Olivine | $LiFePO_4$ | 170/165 | 3.4 | 561 | [50] |
| | $LiMnPO_4$ | 71/168 | 3.8 | 638 | [51] |
| | $LiCoPO_4$ | 167/125 | 4.2 | 525 | [52] |
| Tavorite | $LiFeSO_4F$ | 51/120 | 3.7 | 444 | [53] |
| | $LiVPO_4F$ | 156/129 | 4.2 | 548 | [54] |

**Table 2.**
*Comparison of theoretical and experimental capacity of the major cathode materials and their average operational cell voltage.*

However, the operational cell voltage is relatively lower than the spinel materials and some of the olivine phases. The energy density of $Li_2MnO_3$, NCA, and NMC based phases is higher than other cathode materials. Apparently, it is seen from **Figure 3** that the spinel structure materials should exhibit higher power density than layer and olivine phases since it displays higher ionic diffusivity and conductivity as a function of lithium content as well as temperature. Nonetheless, particle morphologies play a crucial role in the power density and cycling stability of a battery. Three different types of particle morphologies are depicted in **Figure 4** as an example. It should be noted that each type of particle morphology has some inherent advantages and disadvantages for a particular material. It is well known that the spherical dense particle morphology (**Figure 4a**) is beneficial for the long-term cycling of layer structure materials with liquid electrolyte (LE). On the other hand, the nanometer-thick plate-like particle (**Figure 4b**) is good for olivine materials as they exhibit very low ionic and electronic properties transport (see **Figure 3** and **Table 1**). Rodlike particle morphologies (**Figure 4c**) of spinel materials are advantages for achieving high power densities. It is worth mentioning that the operational scenario of ASSB is very different than the conventional LE-based batteries. Unlike LE-based batteries, in SSBs, the solid electrolyte cannot penetrate inside the secondary particles as shown in the schematic (**Figure 4d–f**). Therefore, high power SSB cannot be achieved with a spherical particle in which ionic diffusion length would be longer and all particles might not be completely ionically wired. Submicron cathode particles (single crystals) are highly desirable for high performances SSB. Details are discussed in section 3 for requirements of cathode particles and the fabrication of composite electrodes for high-performance ASSB.

## 2.4 Suitability for solid-state battery

One of the major advantages of SSBs are the safe use of high voltage cathode materials (e. g. LMNO, LRM, and LCoP) which are not feasible in conventional

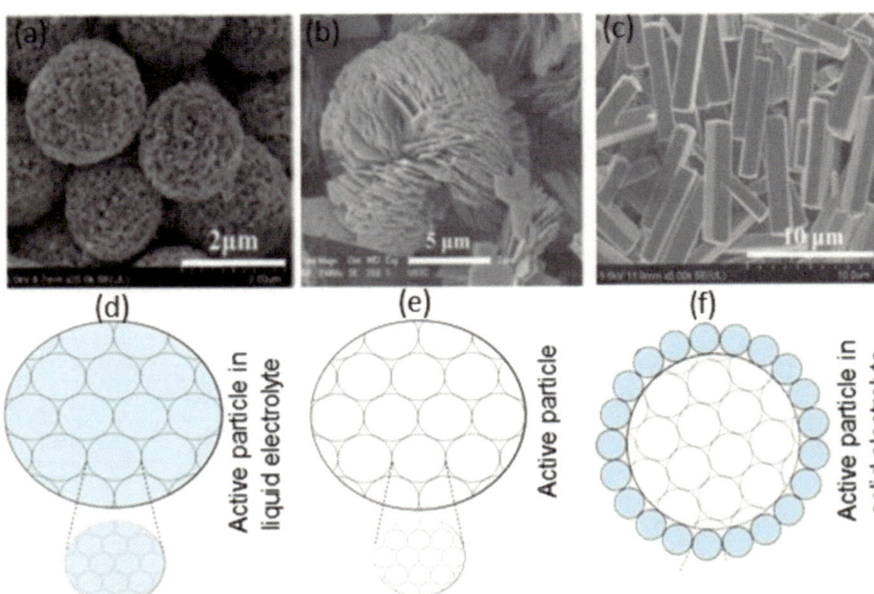


**Figure 4.**
*Particle morphologies of cathode materials (a-c) adapted from the references [55–57] and schematics showing the scenario of liquid and solid electrolytes for ionically wiring of cathode spherical particle (d-f).*

liquid electrolytes due to their limited electrochemical stability window. Thus, SSBs are expected to provide a high energy density at the system level. However, the loose interfacial contact of oxide-based cathode and electrolyte (e. g. NMC/garnet-type electrolyte) creates a severe problem for cell voltage polarization and at room and low temperatures. In addition, a recent report has shown that solid electrolytes (SE) have instability issues while in direct contact with high voltage cathode materials [55]. In such a scenario, one sustainable solution might be the formation of a protective surface coating on cathode materials which would require stabilizing the cathode materials and SE interface by suppressing the possible oxidative reactions in the SSB. Furthermore, the primary cathode particles in the form of single crystals would be more beneficial for SSB than the secondary particles either in spherical or rod-like shapes. Unlike primary nanoparticles, the sphere of secondary particle has many separations or semi-separation regions between neighboring primary particles which is detrimental for SSB operation. Thus, single crystal cathodes are advantageous for use in SSBs because of their good crystallinity, high reaction homogeneity, mechanical strength, and better structural and thermal stability. All these salient features can remarkably improve the electrochemical performance and safety of the SSB.

## 3. Anode materials and designs for SSB

Anode materials can be broadly classified into three major types based on the mechanism of ion storage and electrochemical reactions occurring within the material [58]. The most common and prevalent type of anode material is the intercalation anode (**Figure 5a**) [59–61]. These materials typically possess layered structure into which Li-ion can reversibly insert (intercalate) during cycling of the battery [62]. Graphite, like several other materials (viz. LTO [63, 64], TNO [65]) is an intercalation type anode material. Conventional Li-ion batteries employ graphite as the anode material for hosting Li- ions for reversible intercalation and storage of electrochemical energy. Graphite has a theoretical capacity of 372 mAh $g^{-1}$ which is higher than most cathode materials making it suitable as an anode material [66]. Graphite has demonstrated high coulombic efficiency and cycling performance making it ubiquitous in secondary lithium-ion batteries These materials typically possess lower theoretical capacities; however, they are generally more stable and efficient electrodes. Alternate anode materials can be of deposition or conversion type depending on whether the mobile ion is depositing directly as a metal or as an alloy of a component respectively [3, 58, 67]. Alkali metals (Li, Na, etc.) are examples of deposition type anodes and they possess high theoretical capacity and relatively lower redox potentials [68]. Conversion type materials typically for alloys with the mobile ion (viz. In, Se, Si, etc.) and these also possess high theoretical capacity [69]. The major drawback for the deposition and conversion type anode materials are the electro-chemo-mechanical stability which makes them harder to integrate into functional devices compared to intercalation-type anode materials. Overall, typical reactions for each anode type can be given as:

$$\text{Intercalation Reaction} : \text{Li} + 6\ \text{C} \rightarrow \text{LiC6}\ \left(\text{Capacity} \sim 372\ \text{mAh g}^{-1}\right) \tag{1}$$

$$\text{Deposition Reaction} : \text{Li}^{+} + \text{e}^{-} \rightarrow \text{Li}\ \left(\text{Capacity} \sim 3860\ \text{mAh g}^{-1}\right) \tag{2}$$

$$\text{Conversion Reaction} : \text{x Li}^{+} + \text{Si} + \text{x e}^{-} \rightarrow \text{Li}_x\text{Si}\ \left(\text{Capacity} \sim 3579\ \text{mAh g}^{-1}\right) \tag{3}$$

Solid-state batteries rely on transitioning to high-capacity anode materials of the deposition or conversion type in order to achieve the expected improvements in the

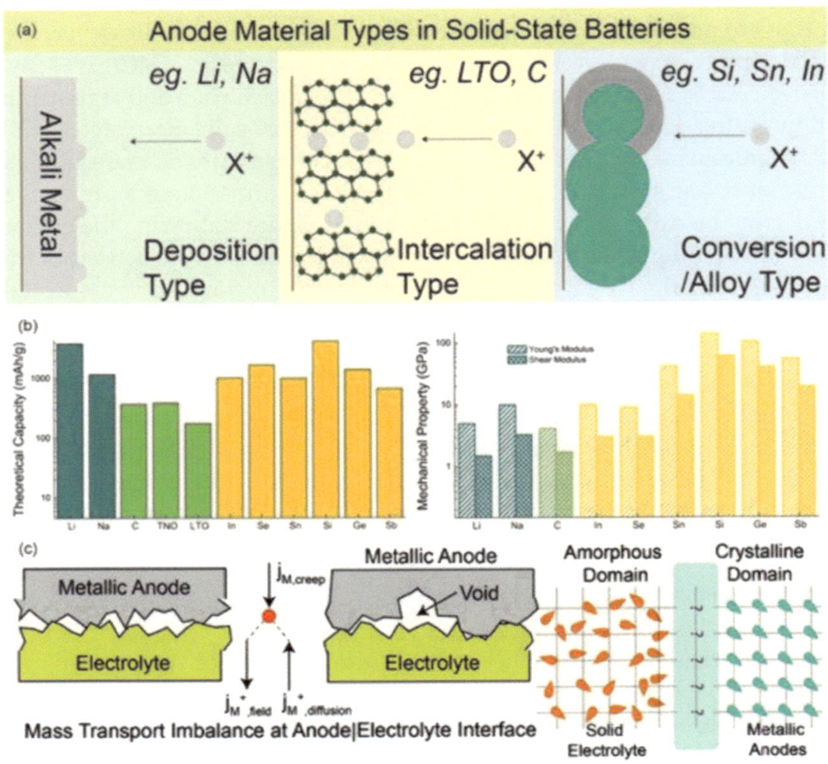


**Figure 5.**
*(a) Schematic diagram showing anode material type and operation mechanism. (b) Nominal capacities and mechanical properties of some common anode materials. (c) Schematic diagram highlighting the challenges with metallic anodes in terms of flux imbalance at the interface of solid electrolyte and the formation of reactive interphase at the electrode | electrolyte boundary.*

energy density. A comparison of nominal capacity of several key deposition, intercalation and conversion type anode materials is provided in **Figure 5b**. It should be noted that the nominal capacities are plotted on a logarithmic scale. The comparison highlights that intercalation anode materials have an order of magnitude lower capacity compared to some conversion/deposition anode materials (375 mAh $g^{-1}$ for graphite; 3860 mAh $g^{-1}$ for Li, 3590 mAh $g^{-1}$ for Si). Mechanical properties of materials are also of key interest for solid-state batteries in order to design solid electrolytes that can mitigate filament growth. Conversion type anode materials typically show higher Young's modulus and shear modulus compared to intercalation and deposition type anodes (**Figure 5b**). Solid electrolyte materials should ideally have shear modulus higher than the anode material in order to mitigate the growth of filaments as proposed by Monroe and Newman [70, 71]. It should be noted that the focus of anode studies with respect to solid-state batteries in the literature is primarily with lithium metal [3]. Relatively fewer reports on intercalation and conversion anode materials are reported and further work is anticipated in these material systems moving ahead.

The key challenges with deposition type anodes, and specifically Li metal will be discussed next. Controlling electrodeposition and electrodissolution morphology for Li metal is imperative to achieving stable solid-state batteries. Specifically, stable morphologies are required at the areal loading of ~5 mAh $cm^{-2}$ of reversible cycling capacity at ~5 mA $cm^{-2}$ plating current density with high coulombic efficiency is far from realization [11]. One major concern with lithium metal is the propensity for filament growth leading to cell failure [72–74]. Filament formation can have significant negative impacts to rate performance, power density and coulombic efficiencies of SSBs. Filament growth typically stems from non-uniform deposition rate at the Li | SE interface. Interfacial kinetics heterogeneity at the Li metal solid electrolyte interface initiates several degradation pathways including filament formation limiting the stability and performance of solid-state batteries. In addition to growth of filaments, high rate electrodissolution from the Li metal can lead to formation of pores that can cause onset of failure [75]. A direct evidence of this was obtained from X-ray tomography measurements of Li | LLZO | Li symmetric cells (**Figure 5c**). Tracking pore evolution during cycling which showed clear cycling behavior (increase in porosity with stripping and decrease in porosity with plating). Mass transport within the Li metal is thus a key challenge and understanding creep and flow behavior of Li is necessary to tune the performance of the system. Interphase formation can also occur during integration of Li metal with solid-electrolytes [15]. Depending on the thermodynamic stability of the solid electrolyte material with lithium metal, three possible interphases can result. These are (i) thermodynamically and kinetically stable (no reaction @ Li | SE interface), (ii) Unstable (unmitigated reaction), and (iii) kinetically metastable (controlled reaction @ Li | SE interface) [3, 68, 73]. With the exception of few materials (viz. LLZO, LiPON), most solid electrolytes undergo reaction with Li metal due to inherent chemical and thermodynamic instability. For some materials, like NASICON-type LAGP and LATP materials as well as LPS thiophosphates, chemical and electrochemical reaction with Li metal leads to an unmitigated growth of an ionically insulating interphase coupled with volume expansion of the material [76–79]. This leads to higher impedances, local stress generation and inhomogeneous current distributions that can cause failure through filament formation, shorting or mechanical fractures. On the other hand, addition of stabilizing agents to the solid electrolyte or introduction of interlayers to these solid electrolytes can lead to formation of a meta-stable interphase that is a mixed ionic and electronic conductor leading to stable solid-state batteries.

Lithium metal stabilization is enabled by several strategies that can be broadly classified into: (i) electrolyte modification [80, 81] (ii) interface modification [82, 83] and (iii) operating parameter modification [84–86]. Electrolyte modification is afforded by additives that can promote the formation of kinetically metastable interphases [73]. For instance, LiI addition to LPS material in conjunction with microstructure control led to improvement of critical current density from $< 0.5$ mA $cm^{-2}$ to $> 4$ mA $cm^{-2}$. Similarly, halide addition to a range of solid electrolytes have shown improved performance in terms of ionic conductivity and critical current density. Interface modification is typically carried out by introducing the use of an interlayer barrier film at the anode | solid electrolyte interface. Atomic layer deposition of materials like $Al_2O_3$, Si, $Li_xAl_{(2-x/3)}O$, $LiXO_3$ (X = Ta, Nb) has shown to improve the performance of lithium metal anodes [87–92]. However, typically the introduction of interlayers is carried out by cost-, time- and equipment- intensive processes that limit the large-scale deployment of such strategies. Another key strategy is modification of operating conditions primarily, temperature and pressure. Indeed, numerous studies have shown the importance of a critical stack pressure in order to mitigate the mass transport limitations within lithium metal by enhancing creep flow at higher pressures (**Figure 6a** and **b**) [86, 93, 96–98]. Overpotential at constant lithium stripping current (0.1 mA $cm^{-2}$) shows reversibly changing overpotentials with modification of the stack pressure. Similarly, overpotential as a function of applied current density shows a reduction of overpotential with increasing stack pressure. Silicon and indium-based anodes also show promising performance (**Figure 6d**) [94, 95]. In summary, anode materials

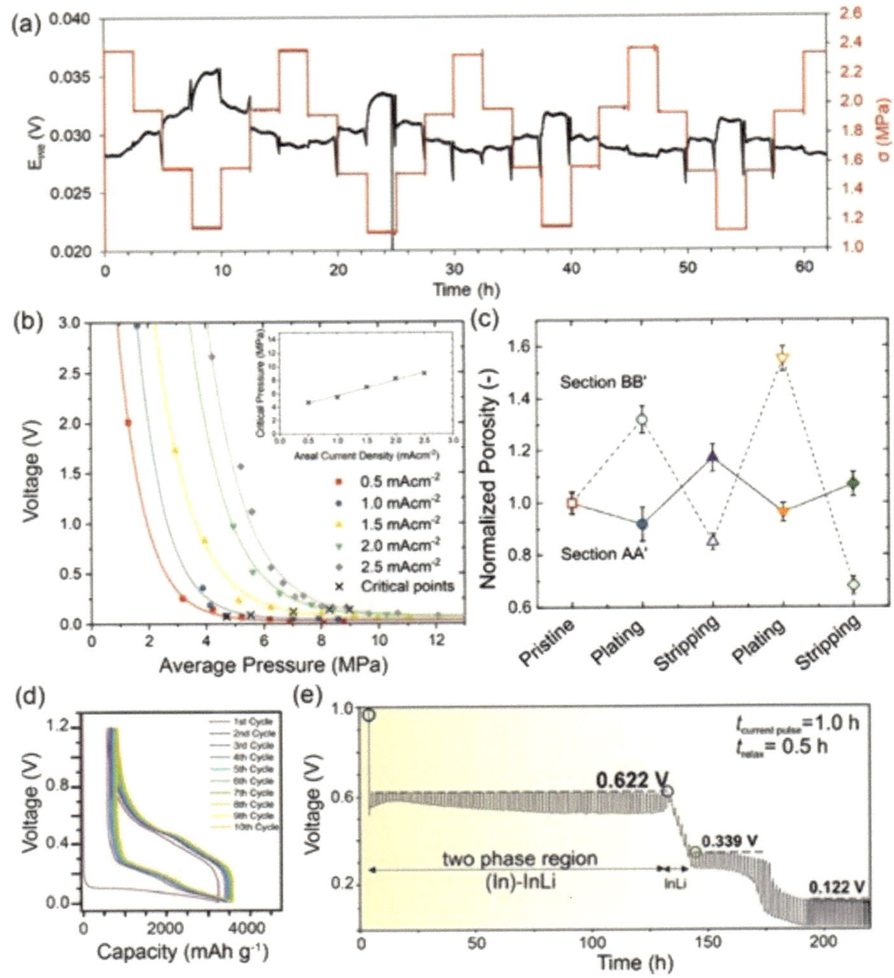


**Figure 6.**
*Summary of key results from anode integration studies in solid state batteries. (a) Potential response of Li | LLZO | Li cell under constant current of 0.1 mA cm2 under varying stack pressures. Reprinted with permission from [84]. (b) Influence of stack pressure on voltage increase for varying current densities for Na | β Alumina | Na cell with the inset showing the critical current density as a function of applied stack pressure. Reprinted with permission from [93]. (c) Porosity for two lithium metal electrodes as a function of cycling steps obtained from X- ray tomography measurements and machine learning segmentation. Reprinted with permission from [75]. (d) Silicon | LPS | Li cell cycling behavior. Silicon particles are spray coated on steel current collectors. Areal loading and current density for the test were 55 μg cm2 and 0.06 mA cm2 respectively. Reprinted with permission from [94]. (e) Potential profile during pulsed lithiation on In metal at 0.2 mA cm2. LPS was used as the solid electrolyte. Reprinted with permission from [95].*

for solid-state batteries need to provide high capacity with high-rate capabilities. Further work on stabilization of anodes under these conditions and demonstration of scalable integration approaches is required for deployment.

## 4. Solid electrolytes

Solid ion conductors have been synthesized in a wide range of chemistries (**Figure 7**) [99]. Currently, most promising electrolyte that have been investigated thoroughly are NASICON (LATP, LAGP) [100], Garnets (LLZO) [101] and Sulfides (LPS) [102]. However, each of these electrolytes have distinct limitations which are hindering their deployment in ASSBs. NASICONs have been widely investigated

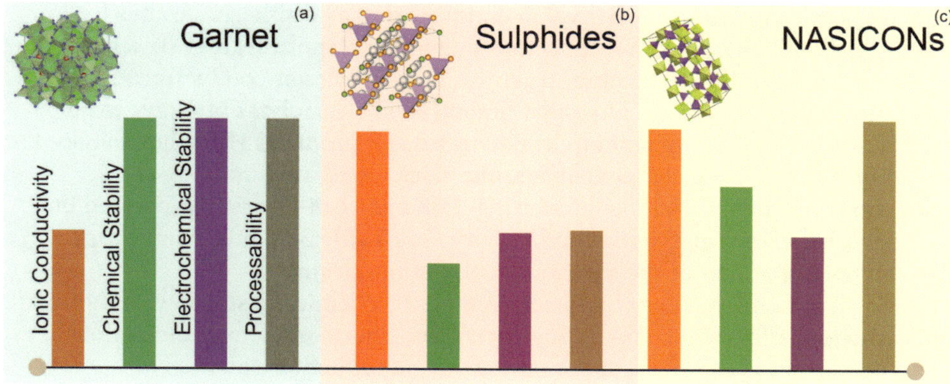


**Figure 7.**
*Schematic diagram highlighting the differences in properties of three major classes of solid electrolytes: (a) Garnet, (b) Sulphides, and (c) NASICONs.*

for not only Na-ion but also Li-ion all-solid-state batteries [103, 104]. $Li_{1.3}Al_{0.3}Ti_{1.7}(PO_4)_3$ (LATP) and $Li_{1.5}Al_{0.5}Ge_{1.5}(PO_4)_3$ (LAGP) are the two most popular NASICON electrolytes for Li-metal solid-state batteries. These electrolytes have high ionic conductivities ($\sim10^{-3}$ S $cm^{-1}$) but suffer from stability issues as $Ti^{4+}$ and $Ge^{4+}$ undergo reduction in contact with Li metal anode [105–107]. First principles studies on these LATP and LAGP materials report operating voltage windows between 2.17 – 4.21V and 2.7 – 4.21V respectively. Garnet type electrolytes have garnered great attention due to their high electrochemical voltage window which enables high power density ASSB and enables long cycling life. However, garnet electrolytes have ionic conductivity one order of magnitude lower than LATP and LAGP. Garnet electrolytes also suffer from environmental instability which makes the processing of garnet type solid electrolyte cost intensive [108]. While a plethora of solid electrolyte classes have been explored over the years, the sodium superionic conductors or the NASICON class are slowly being re-examined for their high ionic conductivities, mechanical robustness and good chemical and electrochemical stabilities [109]. In particular, materials belonging to the NASICON family with phosphate anions are being extensively explored as potential electrolytes and cathode materials for Li, Na, and Mg-ion batteries owing to their high ionic conductivity, thermal and environmental stability [110]. The NASICON type $Na_{1+x}Zr_2Si_xP_{3-x}O_{12}$ ($0 \leq x \leq 3$) is a promising electrolyte material providing high ionic conductivity ($10^{-4}$ S $cm^{-1}$) at room temperature owing to the facile 3D ion conducting pathways. The general formula for these NASICON type materials is $AM1M2(PO_4)_3$ where A can be a monovalent cation $Li^+$, $Na^+$, $K^+$, $Rb^+$, $Cs^+$, $Ag^+$, $Cu^+$, $H^+$, $H_3O^+$, $NH_4^+$, or a divalent cation such as $Mg^{2+}$, $Ca^{2+}$, $Sr^{2+}$, $Ba^{2+}$, $Pb^{2+}$, $Cd^{2+}$, $Zn^{2+}$, $Mn^{2+}$, $Fe^{2+}$, $Co^{2+}$, $Ni^{2+}$ or $Cu^{2+}$ or it can also be vacant. M1 and M2 can be filled with di-, tri-, tetra- or pentavalent transition metal ions within the boundaries of charge balance. NASICONs can crystallize in three different crystal structures, based on the synthesis method, annealing temperature and choice of A, M1 and M2 resulting in α, β and γ-NASICON. Of these, γ-NASICON has the highest symmetry with $R\bar{3}C$ space group which is highly suitable for achieving high ionic conductivities. It is important to explore the possibility of Na-based NASICON materials to be able to conduct Li-ions as well.

Solid electrolytes have limited ionic conductivity at atmospheric temperatures which inhibits the rate capability of ASSB for practical applications (see **Table 3**). Ion transport in polymer electrolyte happens by the complexation of oxides while in inorganic electrolytes, it happens across the crystal lattice sites [111, 112]. Ion transport depends on the available lattice sites and activation barrier for hoping from one lattice to other. Investigation on increasing available lattice sites for ion transport should be carried out for improving ion-transport. Garnet structures can be synthesized for different cubic structures to improve ion-conductivity. Sulfides have shown highest ionic conductivity among solid electrolytes but have major issues with sensitivity ambient environment since it produces $H_2S$ when exposed to humidity. Low bending stiffness of ceramic electrolytes are hindering the processing with roll-to-roll manufacturing. Polymers can be easily processed but they don't have very high ionic conductivity (see **Table 3**). Thus, composite electrolytes can be one of the solutions to this conundrum.

Solid electrolyte must be chemically and electrochemically stable with the anode and cathode material at the operating potentials. Thermodynamic calculations based on the phase equilibria at the oxidation and reduction potential of

| Type | Materials | Conductivity (S $cm^{-1}$) | Advantages | Disadvantages |
|---|---|---|---|---|
| Oxide | Perovskite $Li_{3.3}La_{0.56}TiO_3$ NASICON $LiTi_2(PO_4)_3$ LISICON $Li_{14}Zn(GeO_4)_4$ And garnet $Li_7La_3Zr_3O_{12}$ | $10^{-5}$ to $10^{-3}$ | • High chemical and electrochemical stability<br>• High mechanical strength<br>• High electrochemical oxidation voltage | • Non-flexible<br>• Expensive large-scale production |
| Sulfide | $Li_2S$-$P_2S_5$ $Li_2S$-$P_2S_5$-$MS_x$ | $10^{-7}$ to $10^{-3}$ | • High conductivity<br>• Good mechanical strength and mechanical flexibility<br>• Low grain-boundary resistance | • Low oxidation stability<br>• Sensitive to moisture<br>• Poor compatibility with cathode material |
| Hydride | $LiBH_4$, $LiBH_4$-LiX (X = Cl, Br or I), $LiBH_4$-$LiNH_2$, $LiNH_2$ $Li_3AlH_6$ and $Li_2NH$ | $10^{-7}$ to $10^{-4}$ | • Low grain-boundary resistance<br>• Stable with lithium metal<br>• Good mechanical strength and mechanical flexibility | • Sensitive to moisture<br>• Poor compatibility with cathode materials |
| Halide | Spinel $Li_2ZnI_4$ and antiperovskite $Li_3OCl$ | $10^{-8}$ to $10^{-5}$ | • Stable with lithium metal<br>• Good mechanical strength and mechanical flexibility | • Sensitive to moisture<br>• Low oxidation voltage<br>• Low conductivity |
| Borate and phosphate | $Li_2B_4O_7$, $Li_3PO_4$ and $Li_2O$-$B_2O_3$-$P_2O_5$ | $10^{-7}$ to $10^{-6}$ | • Facile manufacturing process<br>• Good manufacturing reproducibility<br>• Good durability | • Relatively low conductivity |
| Thin film | LiPON | $10^{-6}$ | • Stable with lithium metal<br>• Stable with cathode materials | • Expensive large-scale production |
| Polymer | PEO | $10^{-4}$ | • Stable with lithium metal<br>• Flexible<br>• Easy to produce a large area membrane<br>• Low shear modulus | • Limited thermal stability<br>• Low oxidation voltage (<4 V) |

**Table 3.**
*Solid state electrolyte material for All-Solid-State Lithium Ion batteries.*

| Electrolyte | Reduction potential (vs Li) | Oxidation potential (vs Li) |
|---|---|---|
| Li2S | - | 2.01 |
| LGPS | 1.71 | 2.14 |
| Li3.25Ge0.25P0.75S4 | 1.71 | 2.14 |
| Li3PS4 | 1.71 | 2.31 |
| Li4GeS4 | 1.62 | 2.14 |
| Li7P3S11 | 2.28 | 2.31 |
| Li6PS5Cl | 1.71 | 2.01 |
| Li7P2S8I | 1.71 | 2.31 |
| LiPON | 0.68 | 2.63 |
| LLZO | 0.05 | 2.92 |
| LLTO | 1.75 | 3.71 |
| LATP | 2..17 | 4.21 |
| LAGP | 2.7 | 4.27 |
| LISICON | 1.44 | 3.39 |

**Table 4.**
*Oxidation and reduction potential of well-known solid electrolytes against Li/Li+.*

well-known electrolytes showed that no electrolyte is simultaneously stable at both reductive potential of ~0 V (vs Li) at the negative electrode and at typical positive potential of ~4 V [113]. That's why chemically and electrochemically stable and lithium ion conducting interphase must be formed as depicted in the **Figure 7**. From the first principle calculation, reduction and oxidation potential of well know solid electrolytes are given in the **Table 4**.

## 5. Full cell integration

Material families that can meet ion transport criteria comparable to the state-of-the-art liquid electrolytes have been identified for solid ion conductors. Integration of these materials into a high-performance battery stack is still far from realization. The primary limitation in this regard is the lack of fundamental understanding of the interplay between charge transfer kinetics and mass transport within the system, specifically at the electrode | electrolyte interfaces in addition to other challenges (**Figure 8a**). Typical implementation of lab-scale solid-state batteries is not in traditional coin-cell or pouch-cell formats. Solid-state batteries are typically operated in "pressure cells" that encase the cell system in a container on which a mechanical load is applied (**Figure 8b**) in addition to temperature. Generally, SSBs are reported to function at operating pressures of >100 MPa and elevated temperatures (>50 °C). A quick survey of the reported SSB performance shows that the achieved specific energy and power density of SSBs fall short of required metrics of operations for SSBs of >400 Wh/kg gravimetric energy density and >200 W/kg power density (**Figure 8c**) [11]. Janek et al. have carried out extensive work to understand and decouple the influence of interphase formation and its impact on cycling of SSBs [116, 117]. Typical SSB cycling performance for sulphide based SSBs is depicted in **Figure 8d** [114]. The galvanostatic charge-discharge curves of NMC-811 | Li-In cells with LPS solid electrolyte (separator, catholyte: 30 %, active material: 70 %). SSB cell shows a large first cycle irreversibility (~30 %) compared to an analogous conventional cell (~15 %). Subsequent rate testing shows strong capacity loss at high C-rates with 0 mAh $g^{-1}$ at 1C (**Figure 8d**). Subsequent long-term cycling at 0.1 C shows a strong capacity fade (1-2% each cycle) that is not observed for the conventional cell. The origin of this behavior is identified as a resistive layer formed on the cathode at the high charging voltages which is validated by in situ impedance spectroscopy, SEM and XPS measurements. NCA cathode material with LPSCl solid electrolyte and Li metal anode was investigated in full cells at 5 MPa stack pressure [85]. LNO-coated NCA shows a first cycle irreversibility similar to NMC materials with subsequent cycles showing higher coulombic efficiency (~98%). 80% retention over 200 cycle was observed for this cell at the 5 MPa stack operating pressure and ~3.5 mg $cm^{-2}$ active material loading. The results suggest optimization of the operating conditions (pressure, temperatures) in order to mitigate the formation of filaments and extend SSB lifetimes. Similar studies have been carried out for different cathode and solid electrolyte material combinations that highlight the need of tailoring cathode microstructure, interfaces, reactivity as well as mechanics of the composite cathode. Dixit et al. investigated LFP based cathode composites in conjunction with hybrid solid electrolytes (PEO-LLZO) with varying mechanical properties (**Figure 8e**) [115]. The results indicated that solid electrolyte with higher adhesion properties at the interface shows improved performance due to improved wetting and contact with the cathode. SSB micro-batteries are also investigated as a potential architecture to maximize areal capacity and electrochemically active surface areas for niche applications. $MoS_2$ cathode material in conjunction with PVDF-based solid electrolyte and mesoporous carbon anode was used to fabricate 3D micro-batteries using thin-film coating processes [118]. The results from this study showed improved areal capacity of over an order of

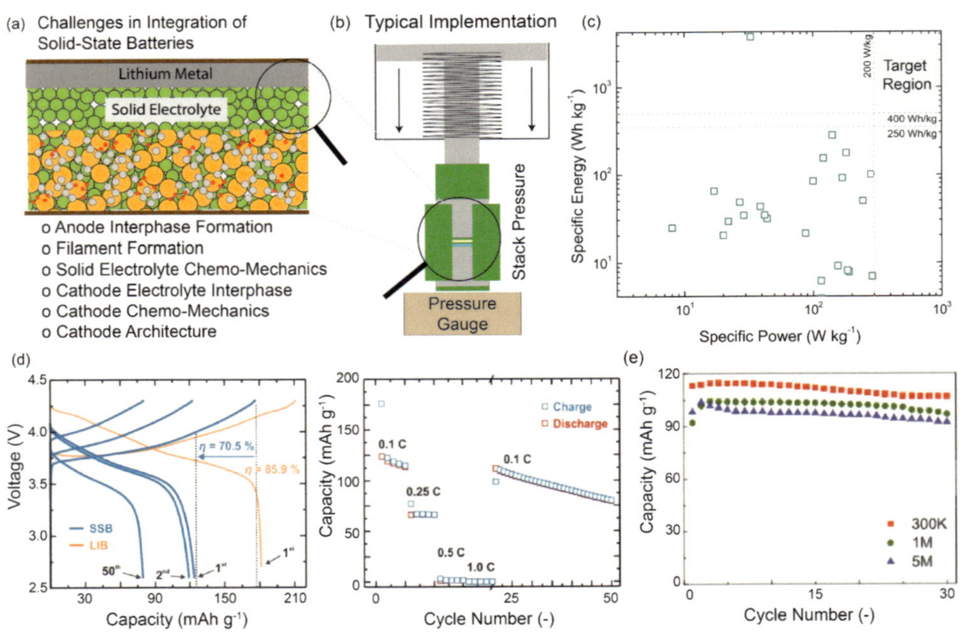


**Figure 8.**
*(a) Schematic diagram highlighting the challenges in integration of all solid-state batteries. (b) Schematic diagram of a typical implementation of a solid-state battery cell. (c) Summary of experimentally reported energy and power density of solid-state batteries. Note that the target performance region for solid state batteries is shaded. (d) Polarization curves and rate performance of NMC-811 / Li-In cells with LPS solid electrolyte (separator, catholyte: 30%, active material: 70%). Reprinted with permission from [114]. (e) Cycling performance of LFP / PEO-LLZO / Li cells for hybrid solid electrolytes with three different molecular weights 300K, 1M and 500 K. Reprinted with permission from [115].*

magnitude for 3D micro-battery compared to a traditional 2D architecture processed identically.

Recently, Dixit et al. carried out a numerical study on investigation on impact of cathode architecture on the energy density of solid-state batteries [14]. They identified a necessity for a large variation in particle size of cathode components in order to achieve higher density composite cathodes as well as to achieve high contact area between the solid-electrolyte and cathode active material. Additionally, the influence of excess anode material to the resultant cell-level energy density was investigated (**Figure 9a–c**). Transitioning to low/no- excess anodes systems can provide significant improvements in terms of cell-level energy density. Dense solid electrolytes (~LLZO) result in high volumetric energy density while low-density solid electrolyte (~PEO) in conjunction with high voltage/capacity cathode materials. Limited demonstrations of completely anode-free cells are observed in literature. Cycling of an in-situ formed Li anode in a NCA | LLZO anode free cell is highlighted here (**Figure 9d** and **e**). The investigated anode free system shows typically low cathode utilization due to unoptimized cathode architecture with highly reversible cycling (coulombic efficiencies ~100%) over 50 cycles [119]. It should be noted that due to changes in "accessible" lithium, certain discharge cycles show higher capacities than the corresponding charge cycle. Another important consideration in solid-state battery architecture is the concept of bipolar stacking. The use of solid electrolyte mitigates the shorting and electrolyte leaking in unit cells allowing for series stacking and reduction of inactive materials in the cell (packaging, sealing, conductor elements). This can lead to improvement in both gravimetric energy density as well power density due to reduction in inactive materials as well as overall resistance of the modules. Initial results with excess-area stainless steel

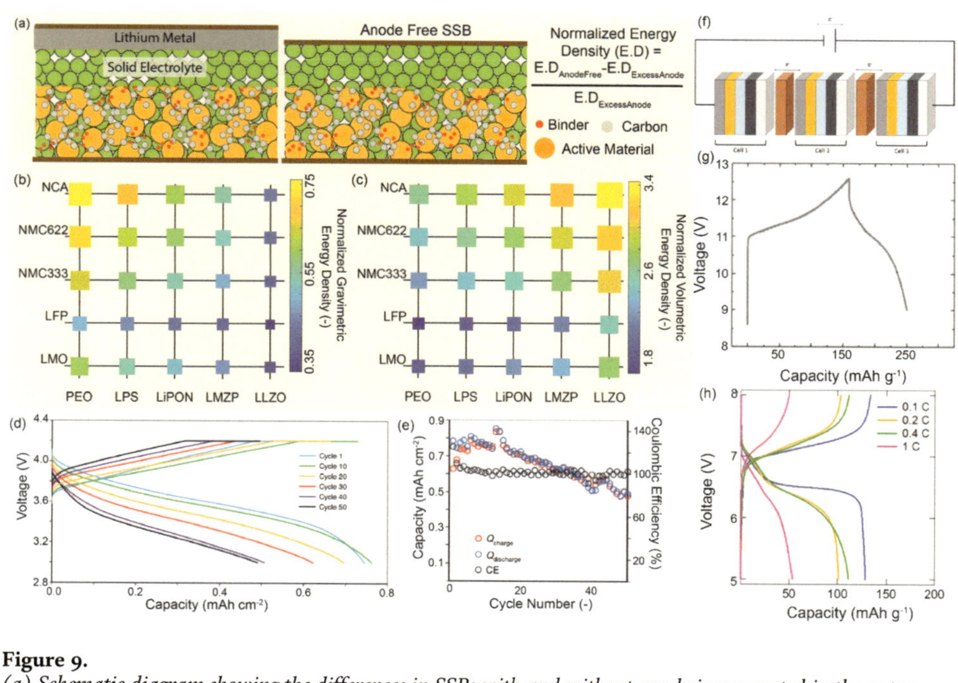


**Figure 9.**
*(a) Schematic diagram showing the differences in SSBs with and without anode incorporated in the system. Effect of transitioning to a no-excess anode system from a 100% excess anode system on (b) gravimetric energy density and (c) volumetric energy density for a range of material combinations. Reprinted with permission from [14]. (d–e) Cycling performance of an anode free NCA/LLZO cell after initial charging cycle at 0.05 mA cm2. Stack pressure of 4 MPa was used for the tests. Reprinted with permission from [119]. (f) Schematic diagram showing bipolar stacking of solid-state batteries. Typical polarization curve for bipolar stacked SSBs with (g) NMC and (h) LFP based cathode materials. Reproduced with permission from [120, 121].*

current collector as a bipolar plate shows promising polarization profiles for NMC622 as well as LFP -based SSBs with polymer based solid electrolytes [120–123]. Subsequent investigations into materials, architectures and cell design for bipolar stacking needs to be carried out for high energy and power density SSBs.

## 6. Perspective and outlook

Battery research at present is at a stage where SSB technology can either redefine the next generation of batteries or fissile out without making an impact. The key difference between the two scenarios is whether the underlying challenges can be systematically mitigated or not. With more commercial enterprises investing heavily into SSB research, there is momentum slowly building to propel and position SSBs as the batteries of the future. Success of SSBs would depend upon the following factors: (i) improving the interfacial issues, (ii) developing unique processing capabilities while minimizing cost, (iii) addressing cell and pack level design challenges when integrating solid state components, and (iv) demonstrating performances exceeding that of advanced lithium-ion batteries. Some potential challenges and opportunities for SSBs include:

With the electric vehicle revolution driving up the demand for batteries that can deliver high energy density, can fast charge, have long cycle and calendar life while maintaining low manufacturing costs, there is definite potential for SSBs to play a major role in achieving these goals. Moreover, the US Department of Energy's target goal for EV batteries include (i) of reducing battery cost to <$100/kWh and ultimately to <$80/kWh, (ii) increasing the range of electric vehicles to 300 miles and (iii) decreasing charge time to 15 minutes or less. A full solid-state battery that can meet these targets that are set for electric vehicles in the next decade is an ambitious endeavor especially when the best anode, cathode, and electrolyte chemistries for such an SSB are not obvious at present.

Urban air mobility is a rapidly pursued avenue that could redefine transportation as we know it. Conventional liquid electrolyte-based batteries are subjected to increasingly stringent safety requirements for use in operation of auxiliary electronics on present day aircraft platforms. Applications with demanding duty cycles such as electric vehicle take off and lift (EVTOL) platforms require batteries that can operate at extreme temperature gradients while being subjected to a multitude of mechanical stresses during operation. SSBs with their safety and energy density advantages are uniquely positioned to be the go-to batteries for such applications. Exploring SSBs for such applications would require coordinated efforts between industrial players and research institutes to initially define the performance requirements for such platforms followed by systematic material and engineering efforts to address the challenges.

i. Extreme environment batteries are a niche yet rapidly growing application due to the increasing public interest into commercial low-earth orbit and interplanetary travel. Certain terrestrial, upper altitude and space applications necessitate secondary energy storage technologies that can function under a very wide modality of extreme environmental conditions including but not limited to temperature, pressure, radiation loads, and mechanical loads. With some present day SSBs operating efficiently at elevated temperatures when compared to the room temperature performance, this presents a unique opportunity for deploying SSBs in these applications. Development of stable, safe, and durable energy storage technologies can have a transformational impact on the application sectors which can include orbital satellites, outer planetary/deep-space probes, land rovers, polar vehicles/end stations, among others.

ii. As an intermediate step, a less-demanding end use application such as SSBs for portable/wearable electronics can be pursued, however, this would risk diverting the focus of the battery R&D community from the biggest projected market for next generation battery systems – electric vehicles and grid storage.

## 7. Conclusion

In summary, despite the challenges, solid-state batteries have great potential for implementation in applications that demand high-energy and safe batteries. Successful deployment of practical SSBs is contingent on addressing the underlying challenges related to materials, processing, and cell engineering. Through this chapter, we have discussed the key issues pertaining to the specific SSB cell components – cathode, solid electrolyte and anode and their interfaces. We envision that our perspectives and outlook discussed in this chapter encourages the readers and inspires solutions that would lead to the eventual practical realization and wide commercial deployment of energy dense solid state battery systems.

## Acknowledgements

This work at the Oak Ridge National Laboratory, managed by UT Battelle, LLC, for the U.S. Department of Energy (DOE) under Contract No. DE-AC05-00-OR22725, was sponsored by the Laboratory Directed Research and Development Program at Oak Ridge National Laboratory.

## Author details

Marm Dixit, Nitin Muralidharan, Anand Parejiya, Ruhul Amin*, Rachid Essehli and Ilias Belharouak*
Electrification and Energy Infrastructure Division, Oak Ridge National Laboratory, Oak Ridge, Tennessee, USA

*Address all correspondence to: aminr@ornl.gov; belharouaki@ornl.gov

## References

[1] The. *Glob. EV Outlook 2020* **2020**.

[2] Janek, et al. *Nat. Energy* **2016**, *1*, 16141.

[3] Hatzell, et al. *ACS Energy Lett.* **2020**, *5*, 922–934.

[4] Wood, et al. *ACS Energy Lett.* **2017**, *2* (3), 664–672.

[5] Xiao, et al. *Nat. Energy* **2020**, *5* (8), 561–568.

[6] Fan, et al. *Adv. Energy Mater.* **2018**, *1702657*, 1–31.

[7] Bachman, et al. *Chem. Rev.* **2016**, *116* (1), 140–162.

[8] Løvvik, et al. 2019.

[9] Mai, et al. *Natl. Renew. Energy Lab.* **2018**, 151.

[10] Kapustin, et al. *Energy Policy* **2020**, *137* (November 2019), 111103.

[11] Randau, et al. *Nat. Energy* **2020**, *5* (3), 259–270.

[12] Shen, et al. *J. Electrochem. Soc.* **2019**, *166* (14), A3182–A3188.

[13] Dixit, et al. *ACS Appl. Mater. Interfaces* **2019**, *11* (48), 45087–45097.

[14] Dixit, et al. *Energy Storage Mater.* **2021**.

[15] Banerjee, et al. *Chem. Rev.* **2020**, *120* (14), 6878–6933.

[16] Wang, et al. *J. Am. Chem. Soc.* **2018**, *140* (1), 250–257.

[17] Whittingham. Materials science and technology; Elsevier Science, 2012.

[18] Thompson. *Phys. Rev. Lett.* **1975**, *35* (26), 1786–1789.

[19] Gamble, et al. *Science (80-. ).* **1971**, *174* (4008), 493–497.

[20] Aronson, et al. *The Journal of Chemical Physics*. NUMBER I I September 18, 1968, pp 470–471.

[21] Whittingham. *Science (80-. ).* **1976**, *192* (4244), 1126–1127.

[22] Dyer, et al. *Science (80-. ).* **2013**, *340* (6134), 847–852.

[23] Islam, et al. *Chemical Society Reviews*. Royal Society of Chemistry January 7, 2014, pp 185–204.

[24] Amin, et al. *Electrochem. Solid-State Lett.* **2007**, *10* (1), A13.

[25] Amin, et al. *J. Electrochem. Soc.* **2016**, *163* (8), A1512–A1517.

[26] Amin, et al. *J. Electrochem. Soc.* **2015**, *162* (7), A1163–A1169.

[27] Amin, et al. *J. Power Sources* **2017**, *348*, 318–325.

[28] Amin, et al. *J. Power Sources* **2017**, *348*, 311–317.

[29] Wu, et al. *J. Electrochem. Soc.* **2012**, *159* (4), A438–A444.

[30] Hao. City University of Hong Kong, 2010.

[31] Seid, et al. *Phys. Chem. Chem. Phys.* **2013**, *15* (45), 19790–19798.

[32] Li, et al. *Energy Environ. Sci.* **2014**, *7* (2), 768–778.

[33] Saadoune, et al. *J. Solid State Chem.* **1998**, *136* (1), 8–15.

[34] Montoro, et al. *Electrochim. Acta* **2004**, *49* (19), 3243–3249.

[35] Sivonxay, et al. *Electrochim. Acta* **2020**, *331*, 135344.

[36] Tritsaris, et al. *J. Phys. Chem. C* **2012**, *116* (42), 22212–22216.

[37] Wu, et al. *Ceram. Int.* **2019**, *45* (4), 5072–5079.

[38] Kim, et al. *Journal of Alloys and Compounds*. Elsevier March 4, 2010, pp L87–L90.

[39] Zhang, et al. *ACS Appl. Mater. Interfaces* **2014**, *6* (20), 17965–17973.

[40] Che, et al. *J. Electrochem. Soc.* **1997**, *144* (12), 4296–4302.

[41] Cho, et al. *Angew. Chemie - Int. Ed.* **2003**, *42* (14), 1618–1621.

[42] Ohzuku, et al. *J. Electrochem. Soc.* **1993**, *140* (7), 1862–1870.

[43] Bruce, et al. In *Journal of Materials Chemistry*; Royal Society of Chemistry, 1999; Vol. 9, pp 193–198.

[44] Tu, et al. *Electrochim. Acta* **2006**, *51* (28), 6456–6462.

[45] Lin, et al. *Nat. Commun.* **2014**, *5* (1), 3529.

[46] Wang, et al. *Adv. Energy Mater.* **2013**, *3* (10), 1358–1367.

[47] Lee, et al. *Nano Lett.* **2014**, *14* (2), 993–999.

[48] Nisar, et al. *J. Power Sources* **2018**, *396*, 774–781.

[49] Choi, et al. *J. Electrochem. Soc.* **2002**, *149* (2), A162.

[50] Yamada, et al. *J. Electrochem. Soc.* **2001**, *148* (3), A224.

[51] Choi, et al. *Nano Lett.* **2010**, *10* (8), 2799–2805.

[52] Lloris, et al. *Electrochem. Solid-State Lett.* **2002**, *5* (10), A234.

[53] Barker, et al. In *Journal of Power Sources*; Elsevier, 2005; Vol. 146, pp 516–520.

[54] Sobkowiak, et al. *Chem. Mater.* **2013**, *25* (15), 3020–3029.

[55] Guo, et al. *Electrochim. Acta* **2017**, *236*, 171–179.

[56] Ma, et al. *Angew. Chemie - Int. Ed.* **2016**, *55* (11), 3667–3671.

[57] Wang, et al. *J. Power Sources* **2011**, *196* (23), 10176–10182.

[58] Puthusseri, et al. *ACS Omega* **2018**, *3* (4), 4591–4601.

[59] Parikh, et al. *J. Electrochem. Soc.* **2019**, *166* (14), A3377–A3383.

[60] Wood, et al. *Energy Storage Mater.* **2020**, *29*, 254–265.

[61] Jaiser, et al. *J. Power Sources* **2016**, *318*, 210–219.

[62] Takami, et al. *J. Electrochem. Soc.* **2000**, *142* (2).

[63] Nordh, et al. *ChemElectroChem* **2017**, *4* (10), 2683–2692.

[64] Pfenninger, et al. *Adv. Funct. Mater.* **2018**, *28* (21), 1–8.

[65] Lee, et al. *Sci. Rep.* **2017**, *7* (1), 1–13.

[66] Asenbauer, et al. *Sustain. Energy Fuels* **2020**, *4* (11), 5387–5416.

[67] El kharbachi, et al. *Solid State Ionics* **2018**, *317* (September 2017), 263–267.

[68] Krauskopf, et al. *Chem. Rev.* **2020**.

[69] Baggetto, et al. *Adv. Funct. Mater.* **2008**, *18* (7), 1057–1066.

[70] Monroe, et al. *J. Electrochem. Soc.* **2003**, *150* (10), A1377–A1384.

[71] Monroe, et al. *J. Electrochem. Soc.* **2005**, *152* (2), A396.

[72] Shen, et al. *ACS Energy Lett.* **2018**, acsenergylett.8b00249.

[73] Dixit, et al. *Matter* **2020**, *3* (6), 2138–2159.

[74] Dixit, et al. *ACS Appl. Mater. Interfaces* **2018**, *11*, 2022–2030.

[75] Dixit, et al. *ACS Appl. Energy Mater.* **2020**, *3* (10), 9534–9542.

[76] Lewis, et al. *ACS Energy Lett.* **2019**, *4*, 591–599.

[77] Albertus, et al. *ACS Energy Lett.* **2021**, 1399–1404.

[78] Tippens, et al. *ACS Energy Lett.* **2019**, *4* (6), 1475–1483.

[79] Lewis, et al. *Nat. Mater.* **2021**, *c*.

[80] Bonnick, et al. *J. Mater. Chem. A* **2019**, *7* (42), 24173–24179.

[81] Singh, et al. *Chem. Mater.* **2020**.

[82] Han, et al. *Nat. Mater.* **2017**, *16* (5), 572–579.

[83] Liu, et al. *ACS Mater. Lett.* **2020**, 665–670.

[84] Wang, et al. *Joule* **2019**, *3* (9), 2165–2178.

[85] Doux, et al. *Adv. Energy Mater.* **2020**, *10* (1).

[86] Kasemchainan, et al. *Nat. Mater.* **2019**.

[87] Han, et al. *Nat. Mater.* **2017**, *16* (5), 572–579.

[88] Luo, et al. *J. Am. Chem. Soc.* **2016**, *138* (37), 12258–12262.

[89] Gellert, et al. *J. Electrochem. Soc.* **2015**, *162* (4), A754–A759.

[90] Xu, et al. *Joule* **2018**, *2* (10), 1991–2015.

[91] Cao, et al. *Matter* **2020**, *3* (1), 57–94.

[92] Cheng, et al. *Chem. Rev.* **2017**, *117* (15), 10403–10473.

[93] Spencer Jolly, et al. *ACS Appl. Mater. Interfaces* **2020**, *12* (1), 678–685.

[94] Ohta, et al. *ACS Appl. Energy Mater.* **2019**, *2* (10), 7005–7008.

[95] Santhosha, et al. *Batter. Supercaps* **2019**, *2* (6), 524–529.

[96] Tu, et al. *Cell Reports Phys. Sci.* **2020**, *1* (7), 100106.

[97] Qingsong, et al. **2019**, 1–38.

[98] Zhang, et al. *Cell Reports Phys. Sci.* **2020**, *1* (2), 100012.

[99] Manthiram, et al. *Nat. Rev. Mater.* **2017**, *2* (4).

[100] Zheng, et al. *J. Power Sources* **2018**, *389* (April), 198–213.

[101] Liu, et al. *J. Power Sources* **2018**, *389* (April), 120–134.

[102] Lau, et al. *Adv. Energy Mater.* **2018**, *8* (27), 1–24.

[103] Gao, et al. *Chem* **2018**, *4* (4), 833–844.

[104] Mertens, et al. *Solid State Ionics* **2017**, *309* (July), 180–186.

[105] Sun, et al. *J. Power Sources* **2020**, *471* (July), 1–9.

[106] Meesala, et al. *J. Phys. Chem. C* **2018**, *122* (26), 14383–14389.

[107] Wang, et al. *Electrochim. Acta* **2020**, *334*, 1–8.

[108] Xia, et al. *J. Am. Ceram. Soc.* **2017**, *100* (7), 2832–2839.

[109] Xia, et al. *ACS Apllied Mater. Interfaces* **2016**, *8* (8), 5335–5342.

[110] Parejiya, et al. *ACS Energy Lett.* **2021**, 429–436.

[111] Famprikis, et al. *Nat. Mater.* **2019**, *18* (12), 1278–1291.

[112] Liu, et al. *Chem. - A Eur. J.* **2018**, 1–15.

[113] Luntz, et al. *J. Phys. Chem. Lett.* **2015**, *6* (22), 4599–4604.

[114] Koerver, et al. *Chem. Mater.* **2017**, *29* (13), 5574–5582.

[115] Dixit, et al. *Joule* **2020**, *4* (1), 207–221.

[116] Neumann, et al. *ACS Appl. Mater. Interfaces* **2020**, *12* (8), 9277–9291.

[117] Schmalzried, et al. *Ber. Bunsmges. Phys. Chem* **1998**, *102*, 127–143.

[118] Oudenhoven, et al. *Adv. Energy Mater.* **2011**, *1* (1), 10–33.

[119] Wang, et al. *Nat. Commun.* **2020**, *11* (1), 1–9.

[120] Homann, et al. *ACS Appl. Energy Mater.* **2020**, *3* (4), 3162–3168.

[121] Matsuo, et al. *Sci. Rep.* **2014**, *4*, 1–5.

[122] Gambe, et al. *Sci. Rep.* **2015**, *5*, 10–13.

[123] Liu, et al. *Adv. Sci.* **2020**, *7* (17), 1–13.

Section 2

# Alternative Energy Storage Techniques

Chapter 5

# Physicochemical Approaches for Thin Film Energy Storage Devices through PVD Techniques

*Ramasamy Velmurugan and Balasubramanian Subramanian*

## Abstract

For the fabrication of thin films, Physical Vapor Deposition (PVD) techniques specified greater contribution than all other deposition techniques. Laser Ablation or Pulsed Laser deposition (PLD) technique is the one of most promising techniques for the fabrication of thin films among all other physical vapor deposition. In particular, flexible thin-film energy storage fabrication PLD plays an important role due to its special parameters such as fine thickness control, partial pressure atmospheric condition, pulsed repetition rate, in-situ annealing and microstructure optimization. Very recently, thin film supercapbatteries have been broadly studied, in which the battery and supercapacitor based electrodes are combined to obtain a high specific power and specific energy density and extended cycle stability. In order to fabricate thin film supercapbatteries, electrodes that have a large potential window, high capacitance, and capacity performance are vastly desired. Thus, the presented chapter represents an important enhancement in the growth of economical and eco-friendly thin flexible supercapbatteries and confirms their potential in sensible applications such as transport electronics devices and other gadgets.

**Keywords:** pulsed laser deposition, thin film supercapbatteries, micro electronics devices, semi solid state electrolytes, volumetric specific capacities

## 1. Introduction

Increasing energy consumption, rising human population and global warming has raised the necessity to progress alternative energy sources and Electrochemical Energy Storage (EES) devices for futuristic necessities. Further, intensifying demand on high-performance EES for portable microelectronic devices and hybrid electric vehicles has designed giant research thrust in the search for a novel diversity of energy storage devices [1–3]. Most of the modern microelectronic are intended to work on EES such as batteries, Supercapacitors and Hybrid Supercapacitors or Supercapbatteries. In particular, small-scale hybrid devices possessions have become vital requirements for diverse insistent purposes such as biomedical devices and portable electronics. With the intent, EES systems have been well-thought-out as an appropriate power sources for innumerable hands-on potential applications owing to the fast charging/discharging rate capability and exceptional stability. Instantaneously, extensive development in EES technology proposes to interest on the electrochemical performance of electrode materials, electrolytes, and strategy of the devices [4–6]. To make specially, the active material should be sort out in a cost-effective manner for receiving high specific energy and specific power at low cost. However, to meet the greater necessities of upcoming systems, Researchers need to expand their performance by designing novel materials with high energy and power density concurrently. In the past few years, widespread activities have been defined to emphasize for the capable and simplistic progressions to fabricate thin, stretchable, and signifigant solid-state flexible batteries and supercapacitors, which are well thought-out as one of the opted candidates for most promising power sources in many of the portable and microelectronic applications [7–9].

The thin film energy storage devices like batteries and supercapacitors for satisfying the energy inevitabilities to balance both power and energy densities. In typical supercapbatteries contain two types of energy storage mechanism in a single device that which explicit pseudo capacitive (Faradaic) nature and other one is battery behavior [10, 11]. For emerging flexible thin film energy storage devices fabrication to form thin film electrodes there are variety of coating methods such as Electrochemical deposition (ED) [12], Physical Vapor Deposition (PVD) [13], Chemical Vapor Deposition (CVD) [14], sol–gel coating method, spray coatings, dip coating and innovative thin film coating systems such as Atomic Layer Deposition (ALD) [15] and Pulsed Laser Deposition (PLD) [16] have been employed in the noticeable arrival of thin flexible electrode assemblies. Frequently, the growth of micro and nanostructure coatings in thin film form are more suitable for flexible energy device applications and the most important benefits as the electrode is binder and conductive free in its structural design. This chapter deals with the electrochemical behavior of vanadium pentoxide ($V_2O_5$) and tungsten trioxide ($WO_3$) thin films using PLD as well as thermal Evaporation technique used as different kind of Flexible thin film energy storage devices such as symmetric Supercapacitor and Supercapbatteries. Author demonstrated Transition metal oxides (TMOs) based thin film electrodes for flexible energy storage system rather than bulk electrodes. This chapter shows the recent influence of the TMO based thin films fabricated through PVD techniques for thin film Supercapacitors / Supercapbatteries. Also an example for anode ($WO_3$) and cathode ($V_2O_5$) which based on the use of massive scale to micro / Nano scale structures to enhance the electrochemical properties of new energy systems with appropriate cost. This approach will be defined and delivered for enlightening device performances with extended cycle life of thin film Supercapacitors / Supercapbatteries based on the principal of electrochemical solid state redox reactions.

### 1.1 Thin film energy storage

The expansion of flexible and portable electronics harmfully demands thin flexible and wearable energy storage devices (ESDs) that preserve both high energy and power density with their greater durability and flexibility to influence a vast wearable energy storage systems. Thus, extensive work have been devoted to emerging various types of flexible, stretchable and portable rechargeable supercapacitors (SCs) and batteries [17, 18]. Plentiful development has been accomplished in terms of thin film electrode material design and flexible device structure along with their electrochemical performance. With new type of ESDs, excluding outdated tests applied on supercapacitors, batteries and now supercapbatteries how to evaluate their "viability" and "portability" growths as a concern. Twisting and extending tests are the most used approaches to validate to the stability of flexible thin and stretchable energy storage devices, respectively [19, 20].

## 1.2 Why thin film energy storage

Since the scalability, a growth of micro electrochemical power sources with thin film structural design opens the approach for powering moderated devices such as electronic chip units, Biomedical implantable devices and credit card/ debit cards, and individual sensors systems. The technology of the thin film is useful for understanding the essential properties of the electrode active materials of energy storage system such as Supercapacitors along with lithium ion batteries (cathodes, anodes and solid state electrolytes) free of polymeric binder and carbonaceous preservative [21, 22]. More importantly in the form of thin film energy storage depends up on some specific features like morphology, size, thickness, pore volume etc., here author report why thin film energy storage device important requirement of society.

i. For making of compact sized devices, synthesis of active electrode materials at a bulk scale may not be appropriate reason of the giving out issues. These demands for the requirement for thin film fabrications, which can simplify the expansion of compact devices and significantly binder less for electrode fabrications.

ii. The Bulk SCs have two main drawbacks that boundary their application for transportable electrical and electronic devices. To begin with, the device manufacture consists of high-cost packing materials and device fabrication techniques to avoid the possible leakage of electrolytes, as most of the organic electrolytes are highly toxic and corrosive nature. Furthermore, it is challenging to construct small and flexible thin SC devices using liquid electrolytes attributable to the packaging problem.

iii. Smooth thin films are highly adhesive, can also be used as a reference material for exploration of the morphology's effect on the performance of electrode active materials, as the dimensions, pore volume, surface area and shape of particles influence the physicochemical properties expressively.

Additional imperative factor for the improved attention on thin-film battery resources is their applicability in micro-Lithium Ion Batteries (LIBs). The micro-scaling of devices is ongoing to compact the sizes of devices in addition to their energy demand, which makes many separate applications practicable, if micro-LIBs can be used for the power supply. These energy storage systems can be useful in different fields, such as biomedical implantable devices, laptops-on-chip, or micrometer-sized sensor systems.

### *1.2.1 Supercapacitors*

Supercapacitors (SCs) have significant attention in past years owing to their high power density, long stability of cycle life and ability to bridge gap of the power and energy density between conventional capacitors, fuel cells and Lithium ion batteries (LIBs) Ragone plot of all kinds of energy storage is displayed in **Figure 1**. SCs retain extremely reversible ion adsorption /desorption on the surface of the electrode, nevertheless suffer with low energy density. An evolution of SC with its advantages of greater power density more than batteries, larger energy density delivers than the conventional capacitors, and exceptional durability, is playing an extraordinary role as a favorable candidate to come across this ongoing demand for high efficient EES and to throw out extended necessity on unsustainable fossil fuels [10, 23, 24].

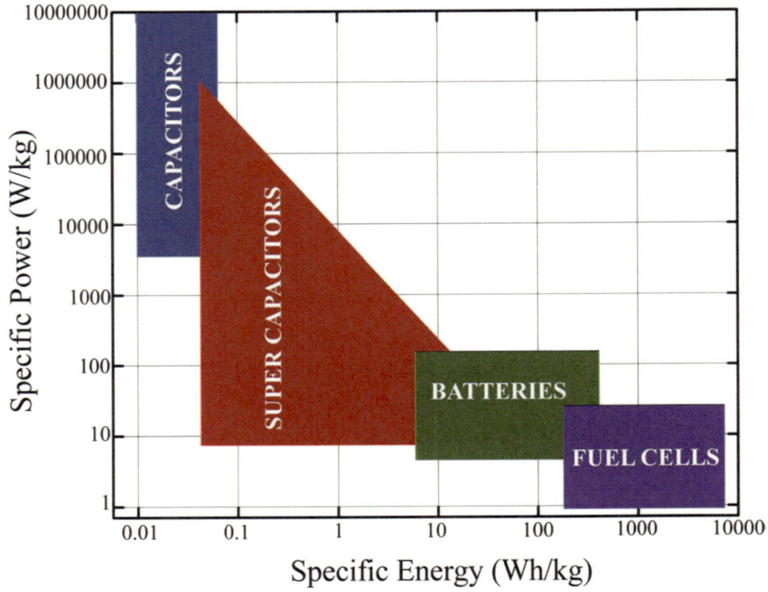


**Figure 1.**
*Ragone plot comparison with all kinds of energy storage system.*

Therefore, a single EES device, which can instantly provide high energy density and high power outputs with long lost, is an extremely desirable.

*1.2.1.1 Symmetric Supercapacitor*

Symmetric supercapacitor is typically assembled by two identical electrodes such as anode and cathode electrodes. The symmetric supercapacitors having limited operating voltage of an aqueous electrolyte up to 1.23 V being restricted by water decomposition, while using organic electrolyte whose voltage window can extend up to 2.7 V. Thin film supercapacitors (TFSCs) have materialized as a new class of electrochemical energy storage device and have considerable attention in recent years. TFSCs make their presence as one of the greatest hopeful energy storage devices attributable to their high power density, outstanding stability, light weight and are easy to handle. Nevertheless, the performance of predictable designs deteriorates extensively as a consequence of electrode and electrolyte exposure to atmosphere along with mechanical distortions for the case of flexible systems [25]. TFSCs are flexible and easily reconfigurable supercapacitors display great potential for application in portable electronics. Moreover, Flexible all-solid state supercapacitors are well-thought-out as a state-of-the art power supply for diminished electrical and electronic devices because they proficiently avoid the leakage of harmful electrolytes, which frequently happens in traditional aqueous electrolyte-based supercapacitors [26, 27]. Numerous challenges limit their applications, such as the thin film composite fabrication process and the underprivileged interfacial compatibility among the electrode and the solid state electrolyte. In contrast to conventional SCs, flexible solid-state SCs have more than a few important benefits containing small size, low weight, exceptional reliability, and an extensive range of practical temperatures. TFSCs hold abundant promise for use as energy storage devices for flexible, stretchable and wearable electronics [7].

Recently the author group reported $V_2O_5$ thin film symmetric SC was fabricated using thermal evaporation technique shown in **Figure 2a**. In this work Ni foam substrate was used as a flexible current collector electrode, Flexible $V_2O_5$ thin film

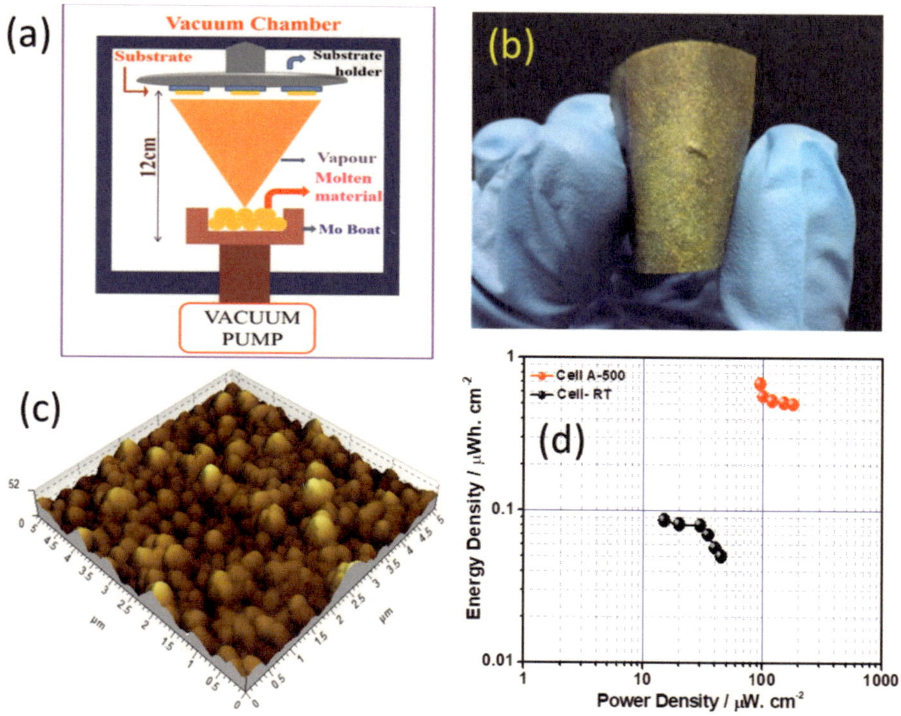


**Figure 2.**
*(a) Schematic diagram of thermal evaporation technique; (b) photographic image $V_2O_5$ thin film annealed at 500°C at CSIR-CECRI, India; (c) AFM 3D topographical morphology of $V_2O_5$ thin film; (d) Ragone plot of $V_2O_5$ symmetric capacitors(Reprinted with permission from Ref. [46]. Copyright 2019 American Chemical Society) .*

electrodes were subjected to observed in a post annealing temperature at 500°C is shown in a **Figure 2b** (photographic image of Ni foam at CECRI, India). The $V_2O_5$ annealed at 500°C thin film was highly conducting nature owing to larger grain size, it is clearly indicated from the Atomic Force Microscopic 3D topographic image as shown in **Figure 2c**. Further author's group compared energy and power density of two symmetric $V_2O_5$ thin film devices such that As-prepared thin film electrode device (Cell-RT) and Annealed at 500°C thin film electrodes device (Cell A-500) is presented in **Figure 2d**. The cell A-500 delivered the maximum areal energy density around 0.7 $\mu Whcm^{-2}$ which is fourteen times greater than as prepared cell-RT (0.05 $\mu Wh\ cm^{-2}$) [28]. Later author's group reported two symmetric thin film SCs using PLD, here this work $V_2O_5$ and $WO_3$ thin film symmetric SCs was fabricated and successfully demonstrated various electrochemical investigation such as Cyclic Voltammogram (CV) and Galvanostatic Charge and Discharge (GCD). The CV curves of both $V_2O_5$ and $WO_3$ symmetric SC devices is exposed **Figure 3a** and **c** reached the maximum voltage up to 1.2 V in a solid state PVA-KOH electrolyte, it is clearly indicated the decomposition appeared each devices above 1.0 V. To avoid this issue, author fixed the voltage window in GCD curve at different current densities of $V_2O_5$ and $WO_3$ thin film symmetric SCs such as 1.0 V and 0.8 V as revealed in **Figure 3c** and **d** [28].

#### *1.2.1.2 Asymmetric Supercapacitor*

Potential window of the symmetric SCs be necessary more or less limitation due to similar materials (same potential widow) used for fabrication, this is one of important difficulty of symmetric SCs. On the way to overwhelm these

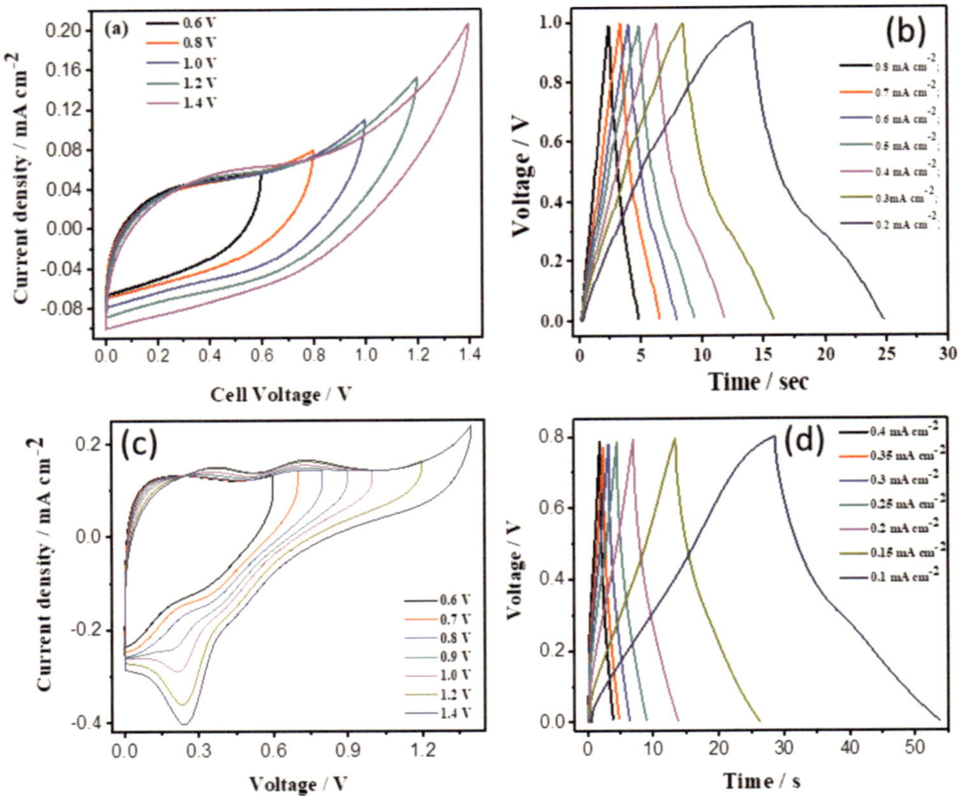


**Figure 3.**
*(a) CV curves $V_2O_5$ symmetric capacitor in different voltage window; (Reprinted with permission from Ref. [46]. Copyright 2019 American Chemical society) (b) GCD curves $V_2O_5$ symmetric capacitor in different current densities; (c) CV curves $WO_3$ symmetric capacitor in different voltage window; (d) GCD curves $V_2O_5$ symmetric capacitor in different current densities (Reprinted with permission from Ref. [28]. Copyright 2020 Royal Society Chemistry).*

issues two dissimilar materials along with different potential widow based active materials are used in device fabrication for extending the voltage window. Asymmetric supercapacitors (ASCs) retain higher theoretical energy density than conventional symmetric SCs have complicated widespread consideration throughout the recent years. Still, there is a huge capacity gap between the two electrodes obviously restrict higher specific energy [29]. Flexible thin film electrodes capacity depends on mass, surface area and thickness of the films, can make the capacity balanced even though optimizing parameters such as weight, volume and thickness of the electrodes. One of the important footnote for several applications, in specific for portable micro electronic devices and hybrid vehicles, the volumetric specific energy is more important than gravimetric specific energy [30, 31].

### *1.2.2 Batteries*

Conventional Li-ion batteries ensuring abound with limitations such that LIB constructed organic electrolytes are highly toxic, corrosive nature and only be handled with glow box atmospheric condition. To avoid this difficulties, solid state batteries (SSB) will be necessary the potential to progress the next generation of energy storage devices over the promises of greater energy density and healthier protection. The main perseverance of solid state electrolyte empowers the predictable of flimsy lithium metal as the anode despite the fact replacing the frequently used inflammable organic electrolyte [32]. Even though the ionic conductivity of definite solid state electrolytes must come together taking place and in some incidents exceeded organic liquid electrolytes, their extensive application has remained inadequate by the excessive interfacial resistance sandwiched between the solid electrolyte and electrode [33, 34].

#### *1.2.2.1 Thin film batteries*

Thin film based LIBs ought to be established loads of wellbeing in consequence of their potential applications as overbearing power sources for micro-electronic devices such as smart cards, sensors and implantable medical devices since many thin film micro-batteries adopt flimsy metal lithium as an anode, development of the cathodes with high energy density becomes significant [35]. All Lithium ion batteries have certain limitations such as spreading out fire, explosive nature of hazards chemicals and overheating at the positive as well as negative electrodes take place while the charge–discharge process in a liquid electrolyte sealed in a metal container [36]. Consequently, all-solid state battery with a solid electrolyte should be very safe and reliable. The schematic stack diagram of solid state thin film battery is shown in **Figure 4** [17]. The thin film SSB consisting anode, cathode and solid state electrolyte in the form of thin film to avoids explosive hazards chemicals, leakage free devices and flexible nature. The electrodes used in thin-film batteries are limited to those that exhibit little volume change during Li ion insertion /deinsertion, since expansion-contraction is restricted in solid-state films [37]. For thin film SSB device fabrication PVD techniques play vital role especially PLD is unique tool for solid state electrolyte deposition in thin film SSB device production. Accordingly, Gil Yoon et al. stated $LiCoO_2$ thin film cathodes fabricated by PLD and the thin film cathode delivered maximum areal capacity 25 μAh $cm^{-2}$ [38]. Kuwata et al. demonstrated solid state electrolyte based $LiCoO_2$ thin film cathodes by PLD and the solid state battery delivered maximum capacity 9.5 μAh $cm^{-2}$ [39]. Park et al. reported Si thin film prepared by PLD for micro battery application, Si thin film electrode delivered maximum areal capacity about 96.7 μAh $cm^{-2}$ [40]. Previously reported literatures reveals that the thin film electrodes used as a coin cell type battery devices. Thus, Author reveals that the thin film based coin cell fabrication by using schematic diagram of thin film battery as displayed in **Figure 5**.

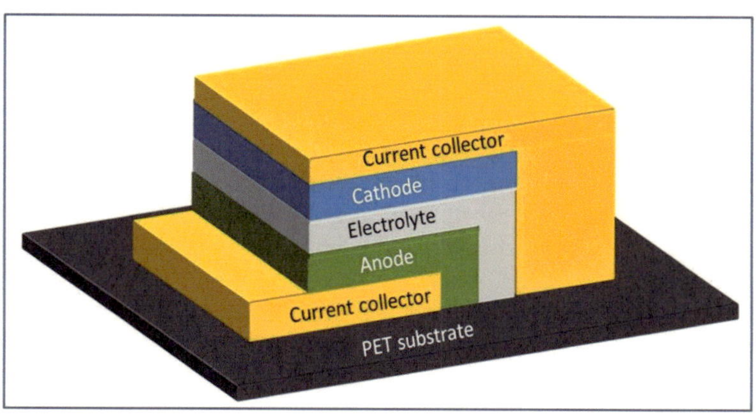


**Figure 4.**
*Schematic stack diagram of solid state thin film battery.*

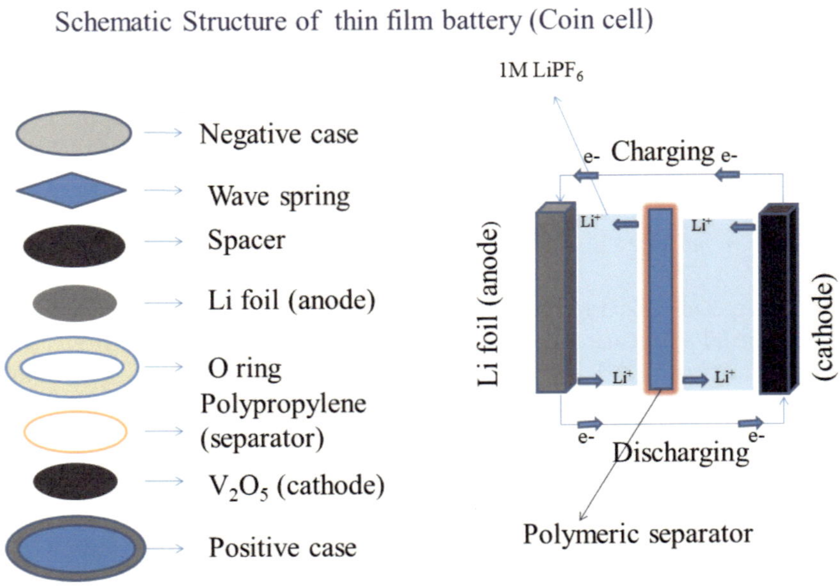


**Figure 5.**
*Schematic representation of structure of thin film battery.*

### *1.2.2.2 From supercapacitor to supercapbatteries*

SCs still have restricted ordinary -life practical application for that their energy density is not comparable to with that of other EESs such as batteries, which the criteria of upcoming energy necessity is far away from adequate to extent. This status spurs ground breaking consequence in the design and preparation of novel hybrid EES that could combining two mechanism is the more advantages than batteries and SCs, which is denoted as supercapbatteries (=supercapacitor + battery) [41]. Therefore, hybrid energy storage devices known as supercapbatteries are rising as a replacement to overwhelm the disadvantage of conventional supercapacitors and batteries, by combining the benefits of each of them, which are superior power and energy density, respectively. A hybrid device is combined by two electrodes with different energy storage mechanism, such as Electric Double Layer Capacitor (EDLC) and faradaic processes; this hybridization of two electrodes could form use of their compatible potential window to increase the voltage window of the device, hence attempt has been made to attain high energy density without yielding constitutional power delivery and very long cycle life of SCs. It deserves that the electrochemical performance of Supercapbattery is nearly attendant to the reasonable design of electrode materials, particularly battery-type materials which deliver large capacity developed from dynamical Faradaic redox reactions. Consequently, the consideration of novel battery-type materials based on various Nanostructures has become a research focal point to encourage the electrochemical performance of Supercapbatteries [42, 43]. Recently Author group designed thin film based supercapbatteries by using PLD. In this work, the fabricated supercapbattery device [28] made by two Transition Metal Oxides (TMOs) such as $WO_3$ and $V_2O_5$, here $WO_3$ exhibited pseudo-capacitive behavior and $V_2O_5$ revealed the battery type behavior. Further, cyclic voltammograms of thin film supercapbattery consisting of $WO_3$ as negative electrode and $V_2O_5$ as positive electrode and their three electrode configuration is presented in **Figure 6a**. The thin film supercapbattery device can reached voltage window 1.8 V (**Figure 6b**) in an aqueous 2 M KOH electrolyte and the thin film device reached 1.6 V in a solid state PVA-KOH gel electrolyte.

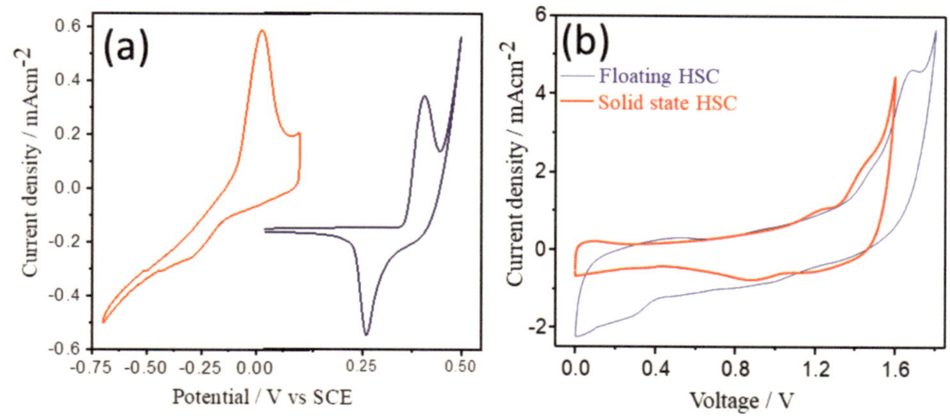


**Figure 6.**
*(a) CV curve combination $V_2O_5$ and $WO_3$ thin films in a three electrode configuration; (b) CV curve comparison of Supercapbattery both aqueous and solid state electrolytes.*

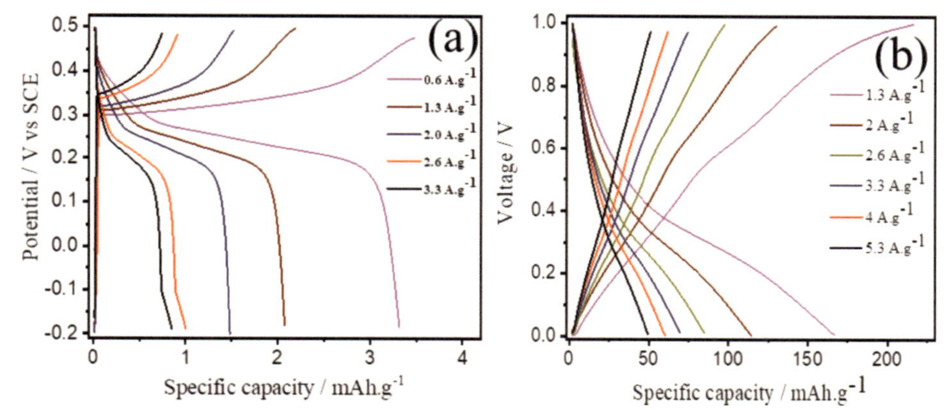


**Figure 7.**
*(a) Charge discharge profile of $V_2O_5$ thin film in a three electrode configuration; (b) charge discharge profile of $V_2O_5$ symmetric supercapacitor.*

### *1.2.2.3 Supercapbatteries in an electrochemical approach*

Supercapbattery devices having high effective battery type electrode materials, which is determined slow kinetics, rate performances quit low and less number of cycling stability. Supercapbattery devices construct the larger potential of battery materials such they are fashionable redox active nature permitting faradaic reaction processes with high energy density materials are appropriate for positive electrodes and pseudo-behavior materials are highly suitable for negative electrodes [41, 43]. In this similarity, author reported the electrochemical investigation of $V_2O_5$ thin film electrode in a three electrode configuration delivered maximum capacity of 3.25 mAh g$^{-1}$ at a current density of 0.6 A g$^{-1}$ as displayed in **Figure 7a**. Even though $V_2O_5$ thin film symmetric device exhibited maximum capacity 160 mAh g$^{-1}$ at a current density of 1.3 Ag$^{-1}$ as shown in **Figure 7b**.

Furthermore, a thin film supercapbattery device was assembled by using PLD process, in this work $V_2O_5$ as a cathode because of it is perform battery nature and $WO_3$ as an anode as it deliver pseudo capacitive behavior. The supercapbattery device shows the better redox behavior in a semi solid state electrolyte was used for fabrication, the thin film device exhibit the maximum voltage of 1.6 V clearly which

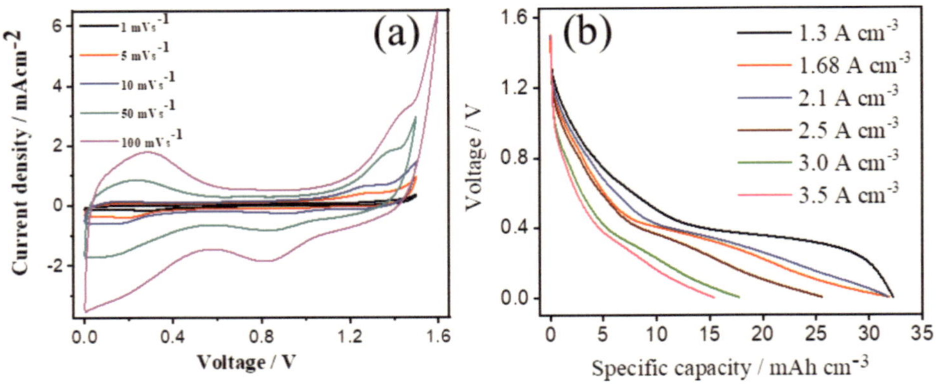


**Figure 8.**
*(a) CV curve of thin film supercapbattery device in different sweep rates; (b) discharge profile for the thin film supercapbattery device.*

indicates CV and discharge profile curves shown in **Figure 8a** and **b**. The supercapbattery device showed excellent rate performance as displayed in **Figure 8b**; the device delivered maximum volumetric discharge capacity of 32 mAh $cm^{-3}$ at a current density of 1.3 A $cm^{-3}$. This is the first thin film supercapbattery energy storage was reported by using PLD system [28]. The agreeing thin film supercapbattery device fabrication cost is very low due to author used alkaline based PVA-KOH electrolyte and the total mass of 0.2 to 0.5 mg of active materials used for thin film supercapbattery fabrication. Therefore, thin film supercapbattery device is economical and eco friendly in nature.

### 1.3 Why PVD techniques for flexible energy storage fabrication

In the past few years ago EES device assembling electrodes such as anodes and cathodes fabrication frequently used approaches like Slurry, Hydrothermal and other synthesis methods ensuring sufficient draw backs for instance the active materials should be very high, low stability owing to require for proper binder, bulky electrodes may not appropriate for micro electronic device fabrication, larger size EES devices, essential proper complex mixture of active materials. To overwhelmed these scenario flexible thin film electrodes fulfill due to less active materials necessary for fabrication for instance compared bulky electrodes 2 to 10 μm thickness thin film electrodes fabrication required active mass of 0.2 to 1 mg, in attendance no necessity of binder required flexible thin film electrode fabrication because thin film electrodes are highly adhesive in nature. Thin film electrodes capable of assembling any miniaturized energy storage devices such as planar Micro- Nano supercapacitors and fiber based energy storages. Intended for emerging flexible thin film energy storage devices, there are numerous thin films coating methods for the occasion of Physical Vapor Deposition (PVD), Chemical Vapor Deposition (CVD), spray coatings, Electrochemical Deposition (ED) and liberal coating methods such as Atomic Layer Deposition (ALD) and Pulsed Laser Deposition (PLD) have been employed in the noticeable appearance of thin flexible electrode fabrications. Frequently, the expansion of micro and nanostructure coatings in thin film form are more suitable for flexible device applications and the major improvement for the thin film electrode is binder and conductive free in its architecture. As we identify that together with the different physical vapor deposition (PVD) techniques such as Thermal evaporation, e-beam evaporation, Magnetron sputtering and PLD, in predominantly PLD is matchless for the intention that its competency to functioning in very high pressure of background reactive gases. The recompenses of the PLD process are flexibility, fine thickness control, high growth rate, quick evaporation and compatible vaporization. PLD also plays an enthusiastic role in influencing the microstructure and phases of the numerous active TMOs and metal sulfides based electrode materials used in the electrode assemblies. In summary, thin film electrode fabrication by PVD techniques are most promising tool to enhance the materials crystalline nature, providing better nano structure with good adhesive properties than films prepared by other techniques.

### *1.3.1 Thermal Evaporation*

Thermal evaporation coating system is a modest technique among all PVD system for thin film fabrication. The schematic diagram of thermal evaporation technique is displayed in **Figure 2a**. In this technique molten material in the form of powder, foils, pellets and salts for thin film fabrication with the help of boats, crucibles and buckets. Usually boats and crucibles made up of molybdenum and tungsten metals owing to they have high meting points. Thermal evaporation technique functioning under the principle of law of conservation such that electrical energy converted to the thermal energy, meanwhile molten materials transferred to one state to another state. Intended for deposition process occurs while applying the current through the boats or crucibles molten material at a particular temperature it goes condensation state to deposit in the form of solid state film on the substrate. In this thermal evaporation technique for a thin film deposition normally used molten material such as some metals foils Al, Cu, Ni, etc. and some metal oxides those materials having low melting points. Recently yen lei et.al approached thermal evaporation technique to form $SnS_2$ thin film for flexible photodetector applications. Ziran Ye et al. reported Ag film on the liquid surface by using thermal evaporation technique for Surface Enhanced Raman scattering (SERS) application [44]. Hailin Hu et al. fabricated Zinc oxide thin film by using this method for Planar Perovskite Solar Cell application [45]. Author group reported $V_2O_5$ thin film supercapacitor prepared by thermal evaporation technique [46]. In conclusion, from the literature thermal evaporation technique is one of the simplest techniques for thin film fabrication for multiple applications also the suitable candidate for thin film energy storage fabrication.

### *1.3.2 Magnetron sputtering*

Thermal evaporation technique has temperature limitation with deposition occurs only for materials having melting point below 1200°C. To overcome this issue Magnetron Sputtering (Schematic diagram is shown in **Figure 9**) coating unit as suitable PVD technique for thin film fabrication attributable to its ensuring more or less special features such as temperature limitations depending on the melting point, thickness control coating unit, easy way to deposit metals, semiconducting materials, ceramic materials and some polymers. In thermal evaporation technique composite thin films cannot be deposition at instant time as it depends on melting point deposition occurs sequentially, In contrast Magnetron sputtering Composite materials deposition happens for instantaneous due to its having multiple target (cathodes) holders. The deposition due to influence of ion bombardment of growing films intensely inspirations their microstructure, and for that reason their physical properties of the film should be changed. Ion bombardment may perhaps intensification to the movement of atoms on the surface of growing film, which effects in increasing the reordering probability of atoms. The thin film process parameters of the magnetron sputtering while deposition is displayed the **Table 1**. All metal oxides, metal sulfides, metal nitrides and metal alloy

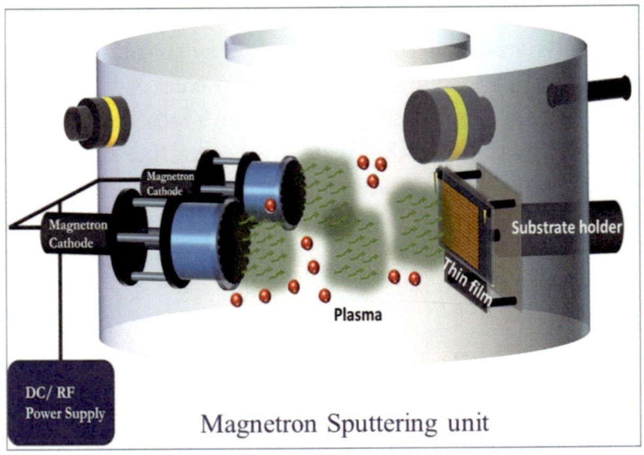


**Figure 9.**
*Schematic diagram of magnetron sputtering unit.*

| Process parameters | Range |
|---|---|
| Temperature | RT – 500 °C |
| Partial pressure | (mbar) $10^{-6}$, $10^{-4}$, $10^{-2}$ |
| Inlet gases | Ar, $N_2$, $O_2$, Acetylene ($C_2H_2$), Methane ($CH_4$) |
| Distance between cathodes and substrate holder | 5 cm |
| Power supply | DC power supply<br>Typically 2000 W<br>RF Power supply<br>Radio frequency: 13.56 MHz<br>Forward power: 0 – 1200 W<br>Reflected power: 0 – 1200 W<br>Biasing<br>-50 V to -1000 V |
| Vacuum pressure | $1 \times 10^{-6}$ (mbar) |
| Substrate rotation Speed | 1 to 10 rpm |
| Target size | 2″ inch & 3″ inch (depending upon cathode) |

**Table 1.**
*Thin film preparation parameters of magnetron sputtering coating unit.*

composites materials based thin films by the promising method to deposit magnetron sputtering for innumerable applications, in particularly more than a few reports available based on energy storage applications. Hyunsik Im et al. reported $CuO_2$ thin film fabricated by using magnetron sputtering for supercapacitor and electro catalyst [47]. Z. Zhang et al. stated molybdenum oxide thin film fabricated via magnetron sputtering for micro supercapacitor application [48]. Zhoucheng Wang et al. demonstrated CrN symmetric thin film supercapacitor with the help of magnetron sputtering unit, and the symmetric device exhibited excellent cycling stability [49]. Zhoucheng Wang et al. studied binder-free titanium nitride thin film electrodes prepared by magnetron sputtering unit for supercapacitors [20]. From the literature magnetron sputtering technology is one of advanced coating system even if comparable lot advantages than thermal evaporation technique, and magnetron sputtering is situated promising tool for flexible thin film electrode fabrications.

### *1.3.3 Pulsed Laser Deposition*

Thermal evaporation and Magnetron sputtering units having few limitations merely two to three composite materials deposited at prompt time. To overwhelmed constraint Laser ablation or PLD (schematic of PLD is shown in **Figure 10a**) had better established to be an unique furthermost suitable techniques for the deposition of thin films comprising an unpredictable through composite stoichiometry. Also PLD has some inimitable advantages such that in-situ temperature controller, partial pressure atmospheric condition, layer by layer coatings, varying the ablation rate exclusively to develop micro/ Nano structured thin film, even this system delivers sufficient microstructure variation and morphologies necessitate for superior electrochemical performance as the most important benefits in PLD are larger deposition rate, precise thickness control unit, capability to functioning in high reactive background gases pressures, and fewer nonconformity from the target composites [50]. The thin film fabrication process parameters of the PLD is exposed in **Table 2**. In this technique Krypton Fluorine (KrF) premixed laser source was used to ablate

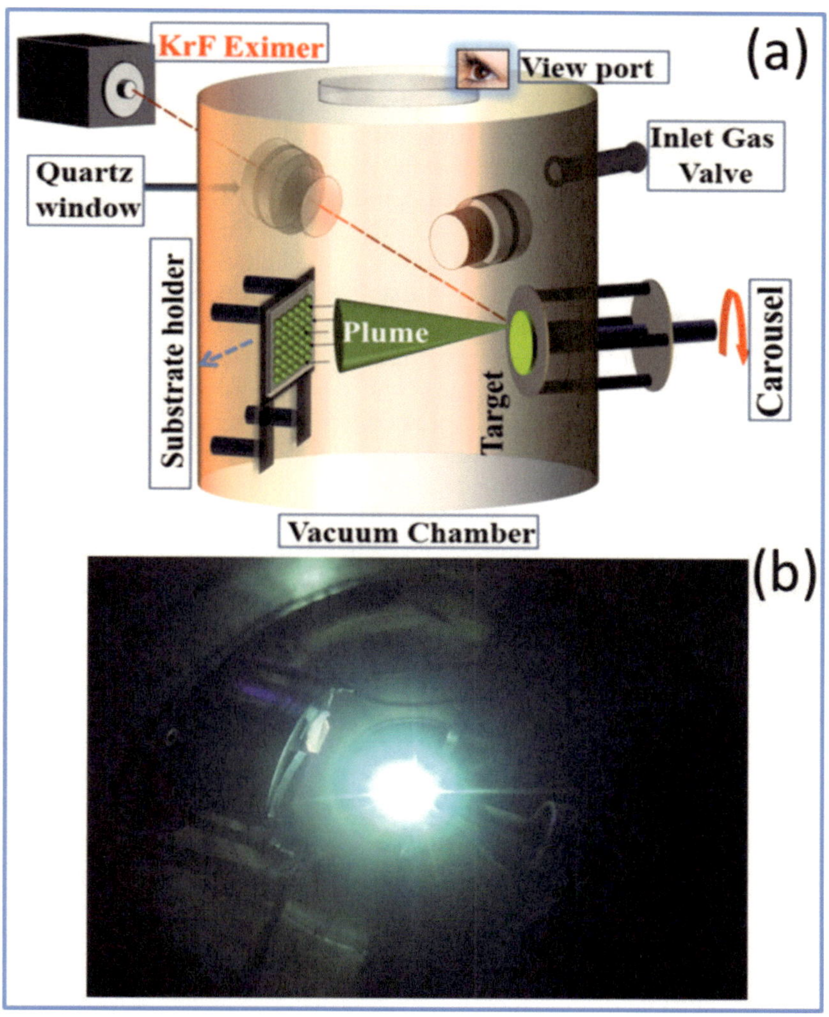


**Figure 10.**
*(a) Schematic diagram of PLD coating unit; (b) photographical image representation for "laser plume" at CSIR-CECRI India (Reprinted with permission from Ref. [28]. Copyright 2020 Royal Society of Chemistry).*

| Process parameters | Range |
|---|---|
| Laser Excimer energy | 100–700 mJ |
| Pulse rate | 1–50 Hz |
| In-situ Temperature | RT –700°C |
| Partial pressure | $10^{-6}$, $10^{-4}$, $10^{-2}$ |
| Inlet gases | Ar, $N_2$, $O_2$ |
| Distance between target and substrate holder | 5 cm |
| Vacuum pressure | $1 \times 10^{-7}$ mbar |
| Target size | 1–2 inches |

**Table 2.**
*Thin film fabrication process parameters in pulsed laser deposition (PLD).*

target of the materials in a high vacuum pressure up to $10^{-7}$ mbar with the help of turbo molecular pump. The laser excimer emits the laser pulse energy 0.8 joule/ pulse at a wavelength 248 nm uses high power (40 W) laser pulses to melt, and evaporate and ionize material from the surface of a target. This laser ablation event produces a high plasma plume that magnify intensely ahead of the target surface, and the produced laser plume is shown in **Figure 10b**. Additionally, PLD unit having rotating target carousel is used to make larger composite materials film in an ambient vacuum condition. PLD is used to fabricate all metals (Au, Pt, Ni, Ag, Cu, Al, etc.), metal oxides ($MnO_2$, $V_2O_5$, $Co_3O_4$, NiO, $SnO_2$ etc.), metal sulfides ($MoS_2$, CoS, NiS, FeS and $VS_2$ etc.), metal nitrides (CrN, TiN, VN and BN), conducting polymers (PANI, PPy etc.), solid state polymers (LIPON etc.) and other metalloid compound thin films for countless applications. While PLD is the stoichiometric conversion of the ablated material on or after the target directed to the substrates and the crystallite phase of the subsequent film is not essentially the similar that the target of materials.

From these consequences PLD is one of the ideal candidates to form micro / Nano structured films for energy storage and energy saving applications. Recently, de Krol et al. fabricated $BiVO_4$ thin film prepared by PLD for solar water splitting application [51]. Wang et al. investigated supercapacitor performances of NiSe thin film electrodes fabricated by PLD technique and the corresponding electrodes delivered specific capacitance value 696 F $g^{-1}$ [52]. Patil et.al studied effect of temperature of $CoFe_2O_4$ thin film prepared via PLD for supercapacitor studies [53]. This work $CoFe_2O$ thin film annealed at 450°C electrode exhibited 777 F $g^{-1}$. Julien et al. examined $Li_2TiO_3$ thin film electrodes produced by PLD aimed at energy storage application. Here the LTO thin film grown at 600°C delivered a specific discharge capacity of 46 μAh $cm^{-2}$ [54]. Lastly Author group demonstrated $WO_3$ and $V_2O_5$ symmetric thin film supercapacitors and Supercapbattery device assembled by using in-situ annealed thin film electrodes prepared by PLD. Thin flexible Supercapbattery device presented superior charge storage performance, also the device displayed high volumetric capacitance about 40 F $cm^{-3}$ [28]. As a final point, PLD is the most appropriate technique for energy storage device fabrication.

### 1.4 Current electrode materials for thin film energy storage

Current commercial flexible energy storage system contains anode and cathode are regularly exclusive based on the intercalation/ deintercalation principal of potassium or lithium ions. Even though these flexible energy storage system by now exhibit a greatly upgraded when compared to the conventional supercapacitors of 10 years ago, their energy storage mechanism principle is also subject to essential limitations prominent to comparably low energy storage system densities. One of the challenging application for supercapbatteries in terms of specific energy and power densities in future portable Micro-electronics. Transition metal oxides such as $RuO_2$, $Fe_2O_3$, $Co_3O_4$, $WO_3$, $V_2O_5$, NiO, $Bi_2O_3$ etc., and ternary metal oxides $NiCo_2O_4$, $ZnCo_2O_4$, $NiMoO_4$, $ZnWO_4$ etc., have long been disregarded as possible electrode materials for all kinds of energy storage system such as Lithium ion batteries, Supercapacitors and Supercapbatteries because of the they having high pore volume with high crystalline nature for insertion / deinsertion of electrolytic ions. The positive electrode as the cathode, the positive electrode frequently has a superior potential than the negative electrode (Anode). The current always streams from the positive electrode to the negative electrode via the peripheral circuit, and the electrons movement in the opposite way. However, cathode (positive) and anode (negative) are well-defined, by the electrochemical electrode reaction being reduction or oxidation.

### *1.4.1 Anodic materials*

For fabrication of hybrid energy storages such as ASCs and Supercapbatteries, anodic materials are promising candidate to meet future energy demands. Usually anodic materials charges stored through an electrolytic ions intercalation/ deintercalation mechanism. As a result, the rate capability performance of hybrid EES is restricted by the sluggish kinetics of ion diffusion in the solid surface, as the surface adsorption–desorption approaches at the cathodic materials are noticeably more rapidly than the Faradaic reactions occurs at the anode, More than a few materials, $Bi_2O_3$, $MoO_3$, $Fe_2O_3$, VN and $WO_3$ are being investigated as the suitable anodes to fabricate hybrid EES because they are having high theoretical specific capacity, faster ions diffusion and easily allowing to intercalation of electrolytic ions.

#### *1.4.1.1 Tungsten trioxide ($WO_3$)*

Tungsten trioxide ($WO_3$) is a noticeable anodic material for the intention that of its low-cost and rich oxidation states ($W^{4+}$, $W^{5+}$, $W^{6+}$); $WO_3$ has in modern times become visible as an apparent anodic electrode material in the development of pseudo-capacitive nature due to its exceptional electrochemical performance and global profusion [55, 56]. However, even though $WO_3$ has well-known its potential as a proficient candidate for a widespread mixture of applications, it's an ideal applicant for thin film EES applications; the active material ought to contain high conducting nature and be capable to providing extraordinary electrochemical performance. Very few of reports on its presentation as an anodic active material in the assembly of a SC in addition to battery necessitate to further investigation in this pathway [57]. Recently author effectively achieved $WO_3$ Nano structure decorated (**Figure 11a**) on the surface of thin films, grown in an in-situ annealed condition by using well established PLD coating unit. Furthermore High resolution transmission electron microscopy (HRTEM) investigation using the $WO_3$ Nano particles with the morphology shown in **Figure 11b** in addition that the elemental distribution analyzes of W and O the color mapping images are shown in **Figure 11c** and **d**. From this contest author revealed that $WO_3$ is the one of the opted anodic material for TFSC device fabrications.

### *1.4.2 Cathodic materials*

Usually, TMOs such as $Co_3O_4$, $MnO_2$, NiO, ZnO, $V_2O_5$, etc. are redox-active behavior and have been used as positive electrode materials for thin flexible energy storage. Most of the TMOs having good electronic conductivity, chemically stable, high theoretical specific capacities, low prices, abundance, and eco-friendly.

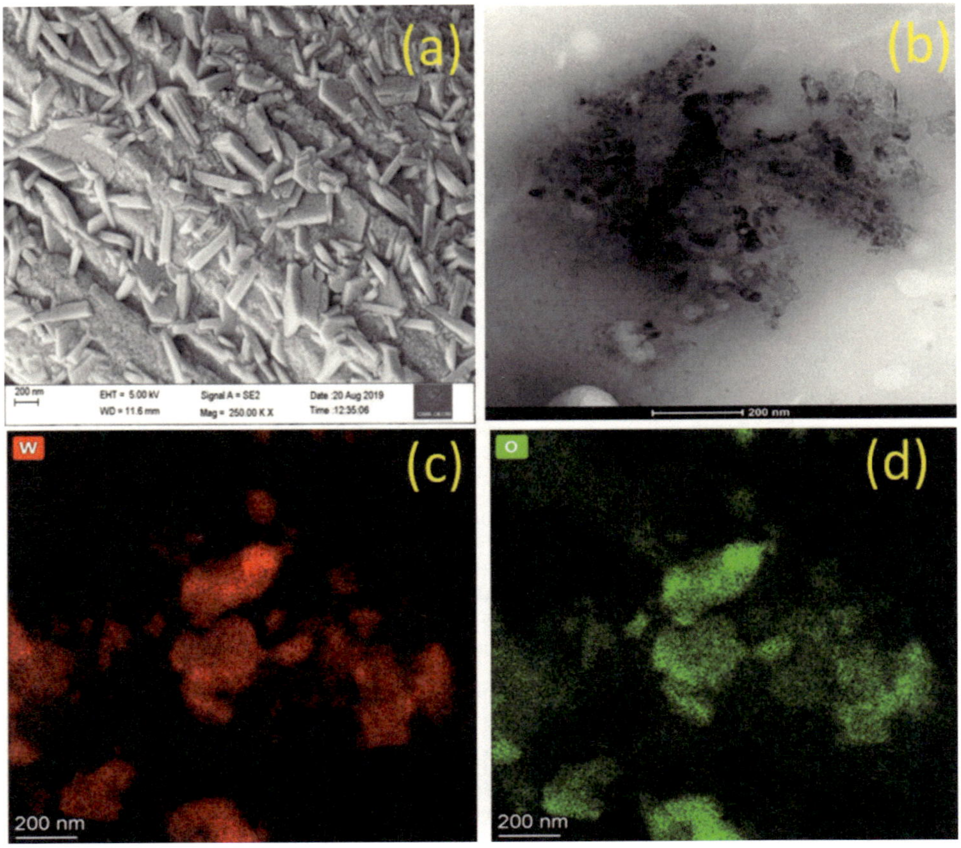


**Figure 11.**
*(a) FESEM morphology of $WO_3$ Nano structures; (b) HRTEM Nano particles image; (c, d) HRTEM- EDAX color mapping images of W and O (Reprinted with permission from Ref. [28]. Copyright 2020 Royal Society of Chemistry).*

### *1.4.2.1 Vanadium pentoxide ($V_2O_5$)*

Vanadium pentoxide ($V_2O_5$) is a well-known electrode active material for EES applications in the middle of vanadium family as stated by Whittingham et., Vanadium pentoxide has variable oxidation states ($V^{5+}$, $V^{4+}$, $V^{3+}$, and $V^{2+}$), permitting it to attain high capacity than the other TMOs and layered structure of $V_2O_5$ creates it highly striking for EES applications [58, 59]. $V_2O_5$ has also paying attention as an active material for improved green EES systems. $V_2O_5$ with diverse morphologies in an adequate particles and thin film Nano structures have been fabricated by a variety of methods. In particularly physical vapor deposition (PVD) techniques are promising tool for thin film Nano structure fabrication. Recently author fabricated $V_2O_5$ thin films by using thermal evaporation technique with different thicknesses such as 210 nm, 380 nm, and 540 nm respectively [46]. As fabricated films further gone to symmetric SC device assembly, further all devices subjected to investigate electrochemical studies. The thin film thickness of 540 nm (cross section **Figure 12b**) symmetric device showed better electrochemical performance as clearly indicated from CV curve shown in **Figure 12a**. Meanwhile, thin film electrodes annealed at 500°C showed redox active behavior than as-prepared film (**Figure 12c**). The post annealing condition is also important for SC device performance because the annealed film morphology (**Figure 12d**) clearly shows the larger grain size.

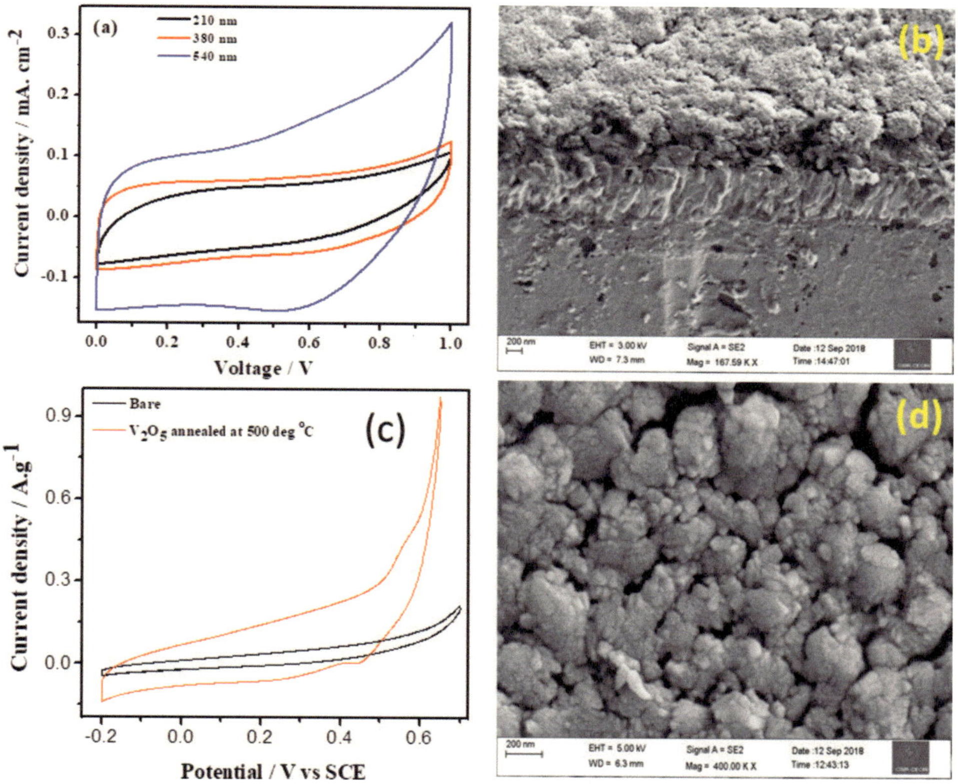


**Figure 12.**
*(a) CV curve comparison of $V_2O_5$ symmetric capacitors in different thicknesses; (b) FESEM cross sectional image of $V_2O_5$ thin film fabricated by thermal evaporation coating unit; (c) CV curve comparison of bare substrate and $V_2O_5$ thin film annealed at 500 °C in a three electrode configuration; (d) FESEM morphology of $V_2O_5$ thin film annealed at 500 °C (Reprinted with permission from Ref. [46]. Copyright 2019 American Chemical Society).*

In thermal evaporation technique, have some draw backs such as large molten materials are required for film fabrication. To overcome this issue author reported $V_2O_5$ thin film electrode fabrication by using PLD. PLD has some unique features such as layer by layer coatings, in-situ annealing condition, fine thickness control and inlet gases atmosphere while film fabrication. The author lastly reported work $V_2O_5$ Nano rods (**Figure 13a**) grown on flexible thin substrate with the help of PLD in an in-situ annealed 500°C at partial pressure atmospheric condition [28]. Further, the Nano structure investigation by using HRTEM is well agreed with Field Emission Scanning Electron Microscope (FESEM) morphology as displayed in **Figure 13b**, also the elemental distribution of vanadium and oxygen was uniformly distributed as presented in **Figure 13c** and **d**.

### 1.5 Flexible electrodes for thin film energy storage

As yet, it is still foremost contest to fabricate flexible thin electrodes with robustness mechanical belongings and outstanding electrochemical performance. The TFSC device fabrication current collector must be an essential tool to supply power to the active materials. Normally, conducting metal foils are used as substrates or electrodes for EEs devices [7, 60]. In particularly, TFSC device manufacture flexible current collectors can be needed; at the present time EES device fabrication usually used flexible electrodes such as 2 dimensional metal foils (Ti foil, Ni foil,

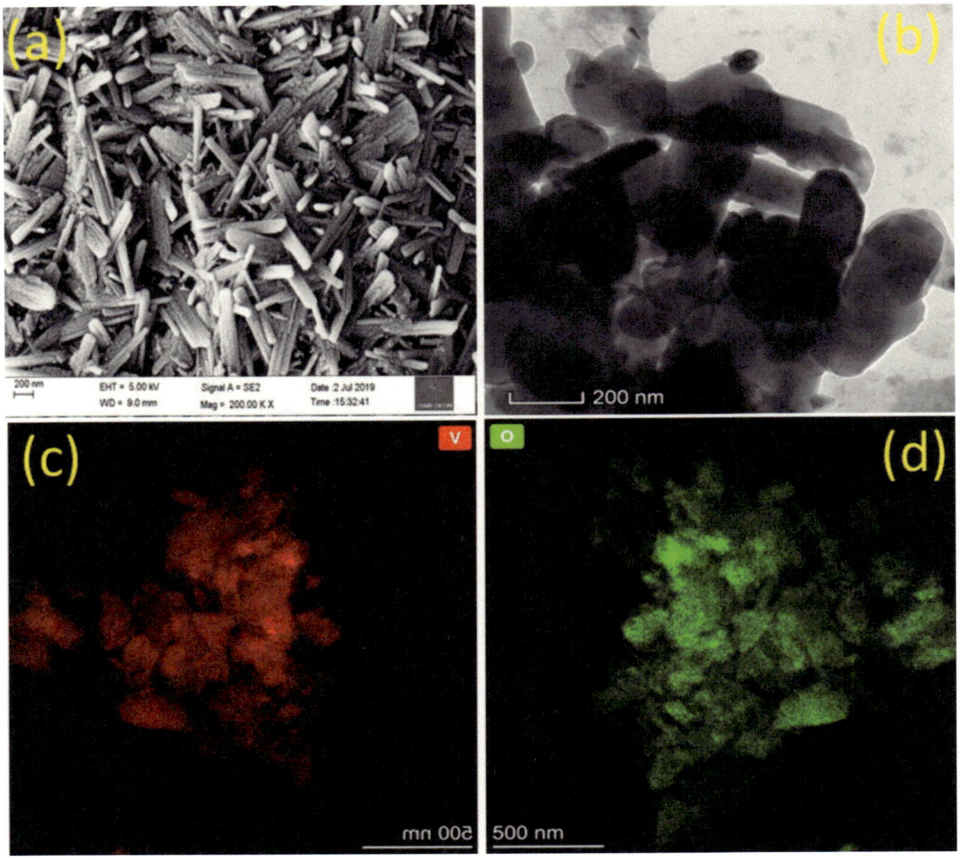


**Figure 13.**
*(a) FESEM morphological image of $V_2O_5$ thin film Nano rods grown by in-situ annealed at 500°C in a partial pressure atmosphere; (b) HRTEM Nano particle morphological image of $V_2O_5$ Nano rods; (c and d) HRTEM-EDAX color mapping images of V and O (Reprinted with permission from Ref. [28]. Copyright 2020 Royal Society of Chemistry).*

and stainless steel foil), conducting carbon clothes, and 3 dimensional arrays (Ni foam, cu foam, and graphite foam) have been widely used for the deposition of a combination of capacitive materials, conducting additives and binder. Nevertheless, metal foils are definitely corroded in aqueous electrolytes, which limits the lifetime of the devices [7, 30]. As a result, foregoing efforts have been attentive on the device design and fabrication of TFSC electrodes by way of non-metal materials. Even though, author used carbon paper substrates in aqueous electrolyte while fabrication of TFSC device used flexible Ni foam array is shown in **Figure 14a** and as prepared TFSC device shown in **Figure 14b**. In set **Figure 14b** clearly indicates Ni foam is one of suitable conducting flexible electrode for TFSC device manufacturing.

## 1.6 Electrolyte for thin film energy storage

The solid-state electrolyte is one of significant key components for fabrication of flexible TFSCs. In assessment to aqueous electrolytes, solid-state electrolytes are at ease to handle, and have superior reliability and an extensive range of working temperature. In addition, with a solid-state electrolyte can avoid a leakage issue, and consequently, which is reducing the device packaging cost [61]. The most extensively used solid-state electrolytes in TFSCs are gel polymeric mixture. A good solid state electrolyte is a non-toxic material, fabrication cost is low and with high ionic conducting nature, excellent stability, functioning at ambient temperature,

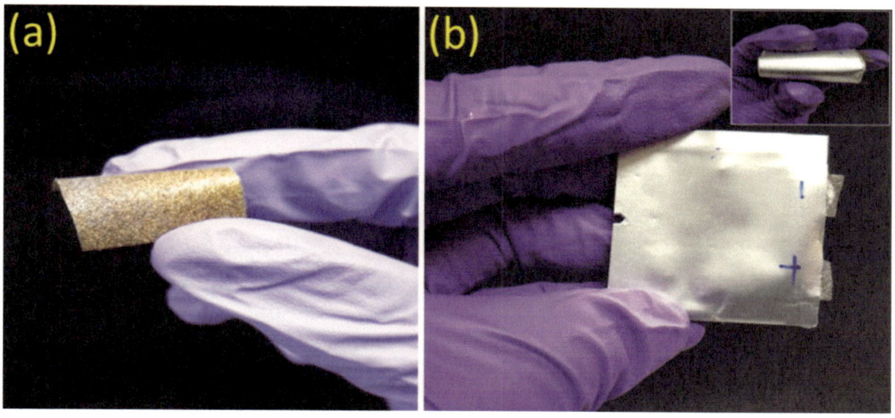


**Figure 14.**
*Photographical image representation at CSIR-CECRI, India (a) $V_2O_5$ thin film deposited on Ni foam substrate; (b) author group fabricated thin film device.*

better mechanical strength and an extensive potential window. In comparison gel polymer electrolytes exhibit superior ionic conductivity than dry solid-polymer electrolytes further down ambient conditions. Gel polymer electrolytes classically contains in a polymeric mixture as the host of an aqueous / organic solvent used as the plasticizer, and a secondary electrolytic salt. Poly ethylene oxide (PEO), poly vinyl alcohol (PVA), polyacrylonitrile (PAN) and poly (methyl methacrylate) (PMMA) are the maximum frequently used for preparing polymeric gel electrolyte mixtures. Author group reported fabrication of TFSC and Supercapbattery devices solid state PVA-KOH gel polymeric mixture was used [28, 46].

### 1.7 Significant parameters for estimating the device performance of flexible energy storage device

There are two significant parameters for estimating performance of Flexible energy storage devices such as volumetric energy density and volumetric power density of a TFSC device can be evaluated by using Eqs. (1) and (2)

$$E = \frac{1}{2}\frac{C_{cell}V^2}{3600}\left(\frac{mWh}{cm^3}\right) \tag{1}$$

$$P = \frac{E}{\Delta t_d} \times 3600\left(\frac{mW}{cm^3}\right) \tag{2}$$

Where $C_{cell}$ is the specific capacitance of the TFSC device, V is the device working voltage and $\Delta t_d$ is the discharge time. Based on Eq. (1), to achieve high volumetric specific energy density and volumetric specific power density, there is a necessity to rise C and V even though reducing $R_s$. Make best use of the TFSC device specific capacitance and voltage window are straight approaches to magnify the volumetric energy density of TFSCs. The working voltage window is determined by the electrode active materials and electrolytes.

The dynamics of thin film solid state battery as well as Supercapbattery devices for estimating specific volumetric capacity from discharge rate performance can be evaluated by using Eq. (3)

$$C_v = \frac{i\Delta t}{3600\,\upsilon}\left(\frac{mAh}{cm^3}\right) \tag{3}$$

Where $C_v$ is specific volumetric capacity of the supercapbattery device. $\upsilon$ is the volume of the thin film supercapbattery device and $\Delta t$ is the discharge time. Author group reported the thin film supercapbattery device showed excellent rate performance and the device delivered maximum volumetric discharge capacity ~32 mAh $cm^{-3}$ at a current density of 1.3 A $cm^{-3}$ [28]. This is unique instance for thin film supercapbattery energy storage was stated via PLD system.

To investigate essentially meaningful volumetric energy and volumetric power densities of a TFSC device, it must be fabricated and examined as a widespread sized and enveloped device. The essential calculation of volumetric energy and volumetric power densities ought to be based on the total area as well as volume of the whole device together with the thin film electrodes, solid-state gel electrolyte, the separator, current collectors and wrapping materials. Author reported supercapbattery device delivered maximum volumetric energy density about 12.5mWh $cm^{-3}$ is displayed in **Figure 15**. Furthermore, the thin film Supercapbattery device delivered the steady performance of cycle stability even if an assorted bending position is shown in **Figure 16a**. Finally, the flexible TFSC tested the practical viability by illuminating Blue Light Emitting Diode (LED) glow (**Figure 16b**) with the series combination thin film devices, TFSCs well thought-out to be probable candidates for use in biomedical and wearable Microelectronic applications.

## 1.8 Reaction kinetic mechanism

The supercapbattery device showed fast kinetics with good storage behavior. The investigated results are extremely specific and exciting in terms of stability, volumetric energy and power density. This development in the supercapbattery device characteristics are essentially attributed to the electrode fabrication where the PLD

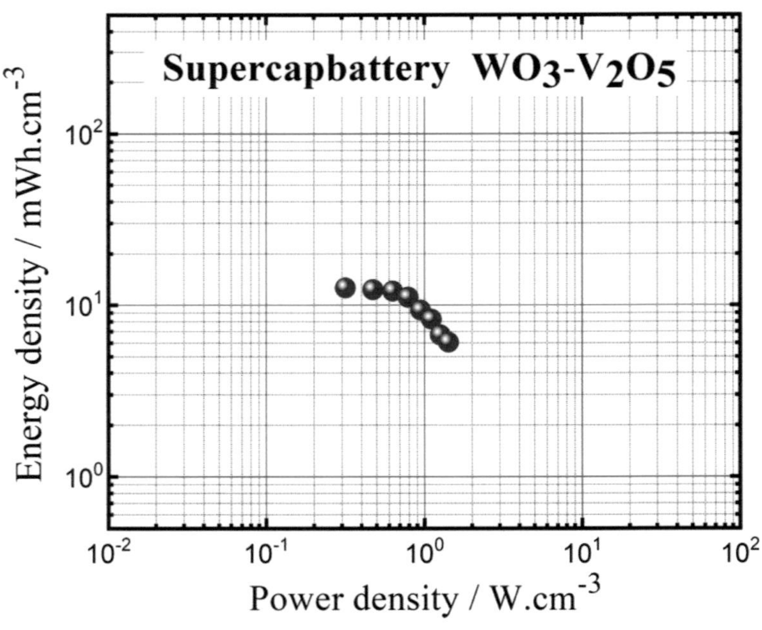


**Figure 15.**
*Ragone plot for thin film supercapbattery device.*

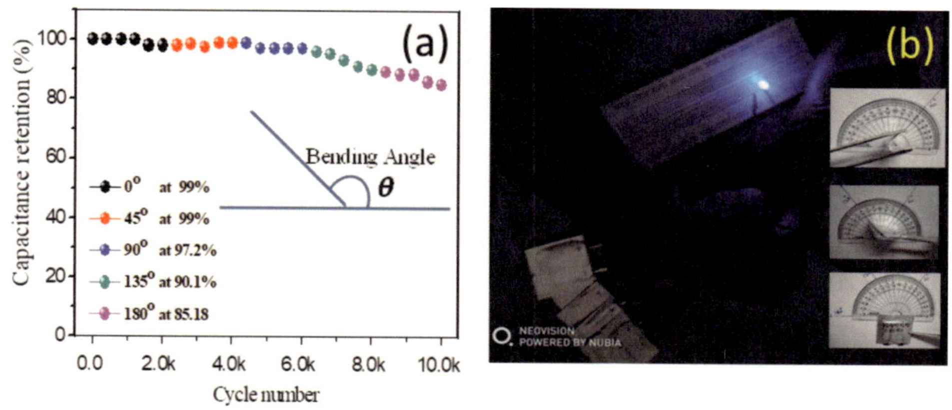


**Figure 16.**
*(a) Stability analysis of supercapbattery for different bent position; (b) photographic image representation for blue LED glow at CSIR-CECRI India (Reprinted with permission from Ref. [46]. Copyright 2019 American Chemical Society).*

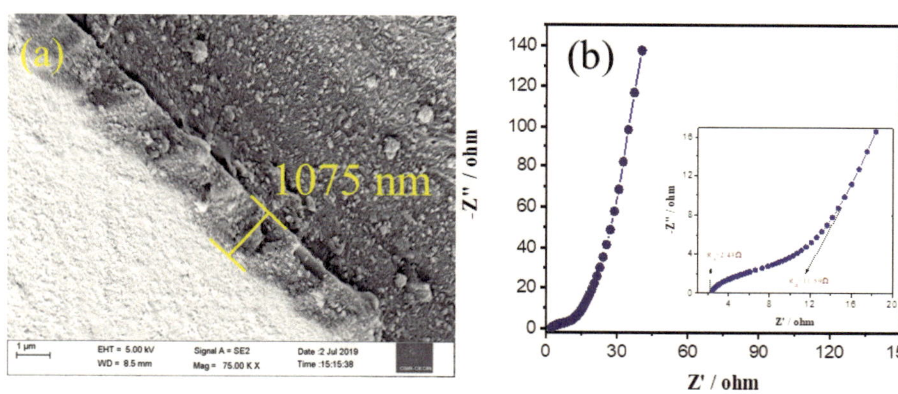


**Figure 17.**
*(a) FESEM cross sectional image for $V_2O_5$ thin film fabricated by PLD; (b) EIS spectra for as fabricated supercapbattery device (Reprinted with permission from Ref. [28]. Copyright 2020 Royal Society of Chemistry).*

deposition process plays an important role in such a Micro/ Nano scale devices. In order to make such supercapbattery device, the charge and mass balancing is very much important to construct, however, it is challenging to balance the charge 100% in practical devices. Instantaneously, in thin film energy storage, balancing of the charge storage can be attained easily by controlling the film fabrication process with the help of advanced coating system. Author's present study, the mass of the thin film electrodes was optimized using the characteristics observed from the three electrode system. On other hand optimized thickness of thin film electrodes are playing very important role for device fabrication, here in author group fabricated thin film electrodes separately with the help of PLD and the thicknesses of $WO_3$ and $V_2O_5$ thin film electrodes such as 1473 nm and 1075 nm is displayed in **Figure 17a**. Further this work reported total thickness of thin film supercapbattery device was 2.5 microns, even if the device presenting good conducting nature, Electrochemical Impedance Spectroscopy (EIS) is the best way to determining Resistance of any electrode or device. The thin film supercapbattery device showed very low charge transfer resistances $R_{ct}$ value 11.9 ohms it's clearly indicating EIS spectra is displayed in **Figure 17b**. Thus the supercapbattery device delivered better electrochemical performances.

### 1.9 Future scope

Nano scale level thin film active materials brought significant improvement for the development of flexible thin film energy storage, Nano complex materials in the form of thin film facilitate accessible of electrolytic ions and an enhance the device rate capability. Nevertheless, an additional side reaction affected by increasing pore area must be taken into consideration for practical wearable and portable electronics. The flexible storage approach to combine in the form of thin film energy storage advantages of different active materials is a hopeful approach for forthcoming development.

Gradually thin film based composite energy storages demands have led to necessities for more specific functions in an electrochemical energy storage devices. Furthermore, outdated Supercapbatteries are undertaking modernizations in different directions to encounter the special necessities of modern society. Here, promising development ways for Supercapbatteries for future as follows

- Microchip energy storage; Easy handling and wearable electronic apparatus is progressively becoming an essential in ordinary life, resulting in the perseverance to improve highly-integrated, diminished and Nano/ Micro-sized energy storage devices. Here, the gradually thin film composite necessities of Nano / micro- scale devices such as smart phones with intellectual operations will necessitate the expansion of materials on the atomic scale in the predictable future

- Portable and self-charging energy storage; Flexible thin wearable and stretchable energy storage devices are foreseeable trend in the imminent development of electrical devices for energy transports, drug delivery, recyclable testing, lighting apparatus, communication equipment's and sensors as well as other applications in which the features of next generation portable products to enable direct wearing or direct connection to skin necessitate thin electrode materials with exceptional flexible, high deformation and low toxicity.

- Implantable energy storage devices; The fast progression of biomedicine and human health witnessing has led to promising demands for implantable very thin energy storage devices to permit for bioinformatics assembly, real-time pathological detection, active drug delivery and clinical usages in which the immeasurable mechanical and kinetic energy from the bio body (e.g. breathing, workout, blood circulation, and heartbeat) can fulfill the self-supply of energy to implantable energy storage devices.

## 2. Conclusion

This chapter converses several properties of thin film influencing their electrochemical performance such as cyclability, energy and power density and so on. Author have performed the comparison studies of two class of energy storage mechanism between supercapacitor and supercapbatteries have been considered to improve large potential window in solid state electrolyte as well as aqueous electrolytes. Flexible thin film supercapbatteries employing with the help of PLD system are expected to exhibit good electrochemical redox activity to deliver high voltage window yet showing a better stability in a post annealed temperature conditions. The thin film supercapbatteries consisting of Tungsten trioxide and vanadium pentoxide has to be a potentially interesting flexible thin film based device, that is simple, low cost, portable and eco friendly. There are no draw backs as the electrode materials because the fabricated electrode materials delivered better cycling stability in a different bent position and both materials having good electrical conductivity. Thus, $WO_3$ and $V_2O_5$ thin film electrodes are promising candidate for flexible thin film energy storage applications and numbers of surveys are highly acceptable to discover the potential of these thin film energy storage materials with highly adhesive film fabrication methods.

## Acknowledgements

Authors would like to acknowledge the Director, CSIR-CECRI, AcSIR Ghaziabad. RV would like to thank Dr. M Ulaganathan shared his fruitful knowledge during TFSC device fabrication. RV would like to thank his funding agency CSIR-HRDG in a CSIR-SRF [File No:30/020(0170)] fellowship scheme.

## List of symbols

| | |
|---|---|
| $R_s$ | Solution resistance (Ω) |
| $R_{ct}$ | charge transfer resistance (Ω) |
| $\Delta t_d$ | discharge time(s) |
| $E_v$ | Volumetric specific energy (mWh $cm^{-3}$) |
| $P_v$ | Volumetric specific power (mW $cm^{-3}$) |
| $E_a$ | Areal energy density (μWh $cm^{-2}$) |
| $P_a$ | Areal power density (μW $cm^{-2}$) |
| $C_s$ | Specific capacitance ($Fg^{-1}$) |
| $C_a$ | Areal capacitance (mF $cm^{-2}$) |
| $C_{cell}$ | Volumetric capacitance (F $cm^{-3}$) |
| V | Voltage window (V) |
| $C_g$ | Specific capacity (mAh $g^{-1}$) |
| $C_V$ | Specific volumetric capacity (mAh $cm^{-3}$) |

## Acronyms and abbreviations

| | |
|---|---|
| PVD | Physical Vapor Deposition |
| PLD | Pulsed Laser deposition |
| EES | Electrochemical Energy Storage |
| ED | Electrochemical deposition |
| CVD | Chemical Vapor Deposition |
| ALD | Atomic Layer Deposition |
| ESDs | Energy storage devices |
| SCs | Supercapacitors |
| LIBs | Lithium Ion Batteries |
| TFSCs | Thin film supercapacitors |
| AFM | Atomic Force Microscopy |
| RT | Room temperature |
| CV | Cyclic Voltammogram |
| GCD | Galvanostatic Charge and Discharge |
| ASCs | Asymmetric supercapacitors |
| SSB | Solid state batteries |
| TMOs | Transition Metal Oxides |
| $V_2O_5$ | Vanadium pentoxide |
| $WO_3$ | Tungsten trioxide |

| | |
|---|---|
| KrF | Krypton Florine |
| FESEM | Field Emission Scanning Electron Microscope |
| HRTEM | High resolution transmission electron microscopy |
| PEO | Poly ethylene oxide |
| PVA | Poly vinyl alcohol |
| PAN | Polyacrylonitrile |
| EIS | Electrochemical Impedance Spectroscopy |
| LED | Light Emitting Diode |
| CSIR | Council of Scientific & Industrial Research |
| CECRI | Central Electrochemical Research Institute |

## Author details

Ramasamy Velmurugan* and Balasubramanian Subramanian
Electroplating and Metal Finishing Division, Central Electrochemical Research Institute (CSIR), Tamilnadu, India

*Address all correspondence to: selvavelanr@gmail.com

## References

[1] Reece R, Lekakou C, Smith PA. A High-Performance Structural Supercapacitor. ACS Appl Mater Interfaces. 2020;12(23):25683-25692.

[2] Libich J, Máca J, Vondrák J, Čech O, Sedlaříková M. Supercapacitors: Properties and applications. J Energy Storage. 2018;17(March):224-227.

[3] Wang Y, Song Y, Xia Y. Electrochemical capacitors: Mechanism, materials, systems, characterization and applications. Chem Soc Rev. 2016;45(21):5925-5950.

[4] Shinde NM, Shinde P V., Mane RS, Ho Kim K. Solution-method processed Bi-type nanoelectrode materials for supercapacitor applications: A review. Renew Sustain Energy Rev. 2021;135(August 2020):110084https://doi.org/10.1016/j.rser.2020.110084

[5] Thind SS, Chang X, Wentzell JS, Chen A. High-performance supercapacitor based on tantalum iridium oxides supported on tungsten oxide nanoplatelets. Electrochem commun. 2016;67:1-5. Available from: http://dx.doi.org/10.1016/j.elecom.2016.03.002

[6] Liu Y, Jiang SP, Shao Z. Intercalation pseudocapacitance in electrochemical energy storage: recent advances in fundamental understanding and materials development. Mater Today Adv 2020;7:100072-100075. Available from: https://doi.org/10.1016/j.mtadv.2020.100072

[7] Lu X, Yu M, Wang G, Tong Y, Li Y. Flexible solid-state supercapacitors: Design, fabrication and applications. Energy Environ Sci. 2014;7(7):2160-2181.

[8] Boruah BD, Misra A. Internal Asymmetric Tandem Supercapacitor for High Working Voltage along with Superior Rate Performance. ACS Energy Lett. 2017;2(8):1720-1728.

[9] Lichchhavi, Lee H, Ohshita Y, Singh AK, Shirage PM. Transformation of Battery to High Performance Pseudocapacitor by the Hybridization of W18O49with RuO2Nanostructures. Langmuir. 2021;37(3):1141-1151.

[10] Chee WK, Lim HN, Zainal Z, Huang NM, Harrison I, Andou Y. Flexible Graphene-Based Supercapacitors: A Review. J Phys Chem C. 2016;120(8):4153-4172.

[11] Lv Z, Li W, Yang L, Loh XJ, Chen X. Custom-Made Electrochemical Energy Storage Devices. ACS Energy Lett. 2019;4(2):606-614.

[12] Wei C, Wu G, Yang S, Liu Q. Electrochemical deposition of layered copper thin films based on the diffusion limited aggregation. Sci Rep. 2016;6(September):1-7.

[13] Rossnagel SM. Thin film deposition with physical vapor deposition and related technologies. J Vac Sci Technol A Vacuum, Surfaces, Film. 2003;21(5):S74–S87.

[14] Cai Z, Liu B, Zou X, Cheng HM. Chemical Vapor Deposition Growth and Applications of Two-Dimensional Materials and Their Heterostructures. Chem Rev. 2018;118(13):6091-6133.

[15] Oviroh PO, Akbarzadeh R, Pan D, Coetzee RAM, Jen TC. New development of atomic layer deposition: processes, methods and applications. Sci Technol Adv Mater. 2019;20(1):465-496. https://doi.org/10.1080/14686996.2019.1599694

[16] Ogugua SN, Ntwaeaborwa OM, Swart HC. Latest development on pulsed laser deposited thin films for advanced luminescence applications. Coatings. 2020;10(11):1-22.

[17] Liang X, Tan F, Wei F, Du J. Research progress of all solid-state thin film lithium Battery. IOP Conf Ser Earth Environ Sci. 2019;218(1):0-15.

[18] Pereira T, Zhanhu Guo, Nieh S, Arias J, Hahn HT. Energy Storage Structural Composites: A Review. J Compos Mater. 2009;43(5):549-560.

[19] Zhao F, Liu W, Qiu T, Gong W Bin, Ma W, Li Q, et al. All Two-Dimensional Pseudocapacitive Sheet Materials for Flexible Asymmetric Solid-State Planar Microsupercapacitors with High Energy Density. ACS Nano. 2020;14(1):603-610.

[20] Wei B, Liang H, Zhang D, Qi Z, Shen H, Wang Z. Magnetron sputtered TiN thin films toward enhanced performance supercapacitor electrodes. Mater Renew Sustain Energy2018;7(2):1-9.https://doi.org/10.1007/s40243-018-0117-9

[21] Chodankar NR, Dubal DP, Gund GS, Lokhande CD. Flexible all-solid-state MnO2 thin films based symmetric supercapacitors. Electrochim Acta. 2015;165:338-347.

[22] Ubale SB, Kale SB, Mane VJ, Bagwade PP, Lokhande CD. SILAR synthesized nanostructured ytterbium sulfide thin film electrodes for symmetric supercapacitors. J Solid State Electrochem. 2021;25(6):1753-1764.

[23] Naskar P, Maiti A, Chakraborty P, Kundu D, Biswas B, Banerjee A. Chemical supercapacitors: a review focusing on metallic compounds and conducting polymers. J Mater Chem A. 2021;9(4):1970-2017.

[24] Zhai Z, Yan W, Dong L, Wang J, Chen C, Lian J, et al. Multi-dimensional materials with layered structures for supercapacitors: Advanced synthesis, supercapacitor performance and functional mechanism. Nano Energy2020;78:105193.https://doi.org/10.1016/j.nanoen.2020.105193

[25] Deng H, Huang J, Hu Z, Chen X, Huang D, Jin T. Fabrication of a Three-Dimensionally Networked MoO3/PPy/rGO Composite for a High-Performance Symmetric Supercapacitor. ACS Omega. 2021;6(14):9426-9432.

[26] Zhou Y, Jin P, Zhou Y, Zhu Y. High-performance symmetric supercapacitors based on carbon nanotube/graphite nanofiber nanocomposites. Sci Rep [Internet]. 2018;8(1):1-7. Available from: http://dx.doi.org/10.1038/s41598-018-27460-8

[27] Zhao C, Zheng W. A review for aqueous electrochemical supercapacitors. Front Energy Res. 2015;3(MAY):1-11.

[28] Velmurugan R, Alagammai P, Ulaganathan M, Subramanian B. High performance: In situ annealed partially pressurized pulsed laser deposited WO3& V2O5thin film electrodes for use as flexible all solid state supercapbatteries. J Mater Chem A. 2020;8(45):24148-24165.

[29] Shinde A V., Chodankar NR, Lokhande VC, Lokhande AC, Ji T, Kim JH, et al. Highly energetic flexible all-solid-state asymmetric supercapacitor with Fe2O3 and CuO thin films. RSC Adv. 2016;6(63):58839-58843.

[30] Chen PC, Shen G, Shi Y, Chen H, Zhou C. Preparation and characterization of flexible asymmetric supercapacitors based on transition-metal-oxide nanowire/single-walled carbon nanotube hybrid thin-film electrodes. ACS Nano. 2010;4(8):4403-4411.

[31] Lv Q, Chi K, Zhang Y, Xiao F, Xiao J, Wang S, et al. Ultrafast charge/discharge solid-state thin-film supercapacitors via regulating the microstructure of transition-metal-oxide. J Mater Chem A. 2017;5(6):2759-2767.

[32] Hayner CM, Zhao X, Kung HH. Materials for rechargeable lithium-ion batteries. Annu Rev Chem Biomol Eng. 2012;3:445-471.

[33] Mukanova A, Jetybayeva A, Myung ST, Kim SS, Bakenov Z. A mini-review on the development of Si-based thin film anodes for Li-ion batteries. Mater Today Energy. 2018;9:49-66. https://doi.org/10.1016/j.mtener.2018.05.004

[34] Marom R, Amalraj SF, Leifer N, Jacob D, Aurbach D. A review of advanced and practical lithium battery materials. J Mater Chem. 2011;21(27):9938-9954.

[35] Moitzheim S, Put B, Vereecken PM. Recent advances in three-dimensional thin-film Li-ion batteries 1991;1-25.

[36] Salah M, Murphy P, Hall C, Francis C, Kerr R, Fabretto M. Pure silicon thin-film anodes for lithium-ion batteries: A review J Power Sources. 2019;414(January):48-67.https://doi.org/10.1016/j.jpowsour.2018.12.068

[37] Qi Z, Wang H. Advanced Thin Film Cathodes for Lithium Ion Batteries. Research. 2020;2020:1-24.

[38] Choi WG, Yoon SG. Structural and electrical properties of LiCoO2 thin-film cathodes deposited on planar and trench structures by liquid-delivery metalorganic chemical vapour deposition. J Power Sources. 2004;125(2):236-241.

[39] Matsuda Y, Kuwata N, Kawamura J. Thin-film lithium batteries with 0.3-30 μm thick LiCoO2 films fabricated by high-rate pulsed laser deposition. Solid State Ionics. 2018;320(February):38-44.

[40] Park MS, Wang GX, Liu HK, Dou SX. Electrochemical properties of Si thin film prepared by pulsed laser deposition for lithium ion micro-batteries. Electrochim Acta. 2006;51(25):5246-5249.

[41] Huang B, Wang H, Liang S, Qin H, Li Y, Luo Z, et al. Two-dimensional porous cobalt–nickel tungstate thin sheets for high performance supercapattery. Energy Storage Mater: 2020;32:105-114. https://doi.org/10.1016/j.ensm.2020.07.014

[42] Chen GZ. Supercapacitor and supercapattery as emerging electrochemical energy stores. Int Mater Rev 2017;62(4):173-202. Available from: http://dx.doi.org/10.1080/09506608.2016.1240914

[43] Chen GZ. Supercapattery: Merit merge of capacitive and Nernstian charge storage mechanisms. Curr Opin Electrochem. 2020;21:358-367.

[44] Sun G, Ye G, Wang K, Lou M, Jia X, Xu F, et al. Deposition of Ag Films on Liquid Substrates via Thermal Evaporation for Surface-Enhanced Raman Scattering. ACS Omega. 2020;5(13):7440-7445.

[45] Rodríguez-Castañeda CA, Hu H, Moreno-Romero PM, Corpus-Mendoza AN, Suárez-Campos G, Miranda-Hernández M, et al. Thermal evaporation–oxidation deposited aluminum oxide as an interfacial modifier to improve the performance and stability of zinc oxide-based planar perovskite solar cells. ACS Appl Energy Mater. 2020;3(10):9618-9627.

[46] Velmurugan R, Premkumar J, Pitchai R, Ulaganathan M, Subramanian B. Robust, flexible, and binder free highly crystalline V2O5 thin film electrodes and their superior supercapacitor performance. ACS Sustain Chem Eng. 2019;7(15):13115-13126.

[47] Pawar SM, Kim J, Inamdar AI, Woo H, Jo Y, Pawar BS, et al. Multi-functional reactively-sputtered copper oxide electrodes for supercapacitor and electro-catalyst in direct methanol fuel

cell applications. Sci Rep [Internet]. 2016;6(October 2015):1-9. Available from: http://dx.doi.org/10.1038/srep21310

[48] Liu C, Li Z, Zhang Z. MoOx thin films deposited by magnetron sputtering as an anode for aqueous micro-supercapacitors. Sci Technol Adv Mater. 2013;14(6).

[49] Wei B, Liang H, Zhang D, Wu Z, Qi Z, Wang Z. CrN thin films prepared by reactive DC magnetron sputtering for symmetric supercapacitors. J Mater Chem A. 2017;5(6):2844-2851.

[50] Santhosh S, Mathankumar M, Selva Chandrasekaran S, Nanda Kumar AK, Murugan P, Subramanian B. Effect of Ablation Rate on the Microstructure and Electrochromic Properties of Pulsed-Laser-Deposited Molybdenum Oxide Thin Films. Langmuir. 2017;33(1):19-33.

[51] Kölbach M, Harbauer K, Ellmer K, Van De Krol R. Elucidating the Pulsed Laser Deposition Process of BiVO4 Photoelectrodes for Solar Water Splitting. J Phys Chem C. 2020;124(8):4438-4447.

[52] Wang Y, Fu ZW, Qin QZ. A nanocrystalline Co3O4 thin film electrode for Li-ion batteries. Thin Solid Films. 2003;441(1-2):19-24.

[53] Nikam SM, Sharma A, Rahaman M, Teli AM, Mujawar SH, Zahn DRT, et al. Pulsed laser deposited CoFe2O4thin films as supercapacitor electrodes. RSC Adv. 2020;10(33):19353-19359.

[54] Lakshmi-Narayana A, Prakash NG, Dhananjaya M, Hussain OM, Jun Qiu Y, Julien CM. Pulsed laser–deposited Li2TiO3 thin film electrodes for energy storage. J Solid State Electrochem. 2020;24(6):1371-1385.

[55] Chang KH, Hu CC, Huang CM, Liu YL, Chang CI. Microwave-assisted hydrothermal synthesis of crystalline WO 3-WO3·0.5H2O mixtures for pseudocapacitors of the asymmetric type. J Power Sources. 2011;196(4):2387-2392. Available from: http://dx.doi.org/10.1016/j.jpowsour.2010.09.078

[56] Zhang G, Guan W, Shen H, Zhang X, Fan W, Lu C, et al. Organic additives-free hydrothermal synthesis and visible-light-driven photodegradation of tetracycline of WO3 nanosheets. Ind Eng Chem Res. 2014;53(13):5443-5450.

[57] Li J, Liu X, Cui J, Sun J. Hydrothermal synthesis of self-assembled hierarchical tungsten oxides hollow spheres and their gas sensing properties. ACS Appl Mater Interfaces. 2015;7(19):10108-10114.

[58] Hu B, Xiang Q, Cen Y, Li S, Liu L, Yu D, et al. In Situ Constructing Flexible V 2 O 5 @GO Composite Thin Film Electrode for Superior Electrochemical Energy Storage . J Electrochem Soc. 2018;165(16):A3738–A3747.

[59] Boruah BD, Nandi S, Misra A. Layered Assembly of Reduced Graphene Oxide and Vanadium Oxide Heterostructure Supercapacitor Electrodes with Larger Surface Area for Efficient Energy-Storage Performance. ACS Appl Energy Mater. 2018;1(4):1567-1574.

[60] Lu X, Zeng Y, Yu M, Zhai T, Liang C, Xie S, et al. Oxygen-deficient hematite nanorods as high-performance and novel negative electrodes for flexible asymmetric supercapacitors. Adv Mater. 2014;26(19):3148-3155.

[61] Yu M, Feng X. Thin-Film Electrode-Based Supercapacitors Joule. 2019;3(2):338-360.https://doi.org/10.1016/j.joule.2018.12.012

Chapter 6

# IPMC Based Flexible Platform: A Boon to the Alternative Energy Solution

*Monojit Mondal, Arkaprava Datta and Tarun Kanti Bhattacharyya*

## Abstract

The ameliorating urge for energy in consonance with the descending environment and attenuation of natural resources leads to the development of alternate energy storage. Realistically, flexible, portable, and lightweight energy storage devices have immense popularity for accessible transportation. In this context, this chapter analyses a possible solution to the problems described aforesaid on IPMC (Ionic Polymer Metal Composite) membranes. Also, this chapter includes porosity induced electrolyte polymer membrane by MCP of Nafion enhances electrical harvesting attribution. The novel and transportable ocean kinetic energy converting platform by IPMC membrane was fabricated and applied for energy conversion. The etching and surface sanding advances the surface area of IPMC to escalate the gas generation rate as an electrolyser. The functionalised infiltrated Nafion nanocomposite membranes are fabricated and analysed for DMFC performance and methanol permeability. Perfluorosulfonic acid polymer electrolyte membranes gained more attention in the former epoch for vast applications in energy, chloro-alkali electrolytes, OER, and polymer electrolyte fuel cells. The direct methanol fuel cell is an excellent alternative to PEFC for managing liquid fuel and higher energy density at low operational temperatures. Nevertheless, polymer electrolyte membranes and direct methanol fuel cells are potential contenders for circulated power and transferable power applications; the substantial technical, scientific, and economic difficulties must be elucidated beforehand commercialisation.

**Keywords:** Energy storage, Fuel cell, IPMC, Flexible storage, smart material, ionic polymer, electro-active polymer, Composite membrane, Electrochemistry

## 1. Introduction

Over the century, the evolution of electronic devices has been rehabilitated rapidly and day by day. The transistor incorporation led to reducing the feature size of any realistic instruments, and besides that, the efficacy has to be ameliorated for that purpose. The continual advancement of the fabrication methodologies of printed circuit boards makes the transistors significantly small, permitting the huge number of transistors in one plate and lessen the consumption of electric power of the integrated circuit. In the present day, many electronic devices are used in miniature sizes and too much power for any operation. Subsequently, energy storage devices are required in the same formation used in those new small devices, batteries, and capacitors, which are hindrances to their improvement. The batteries are analysed, which are lower of flexibility, higher weight, and mainly possess huge space in the particular equipment. Generally, the capacitors with low storage capacity with low energy occupy a huge space, mainly electrolytic capacitors, restrained by shape. Moreover, the capacitors need to solder directly to the integrated circuit; their electrical polarity cannot be reversed in electrolytic capacitors. In this scenario, this chapter explores a probable elucidation to the above-said context based on IPMC (ionic polymer-metal composite) systems. These attain capacitive features after dipping in the ionic solution; then, the membrane becomes flexible and lightweight with stored potential, a new kind of flexible capacitor. Furthermore, the flexible IPMC membrane capacitor is malleable and easily used in polarity reversal based on needs. In most cases, the analysis and application of the IPMC polymer matrix entirely focused on the actuator's application [1–5]. This chapter focused on another part of IPMC, i.e., the energy storage application. The different fabrication procedures and corresponding attributes of the polymer system are discussed thoroughly here. Prominent displacements analyse IPMCs within lower density forces. In the mechanical energy harvesting system, IPMCs are investigated using the generated lower electric power values, but irrespective of using significant mechanical pressures, the reverse function is happened as with the piezoelectric type materials [6, 7]. Present-day, electric double-layer capacitors (EDLCs) and lithium-ion batteries are the maxima used energy storage devices on the scale of important commercial use. Every device exhibits definite attribution and applications that need high energy density ($Whkg^{-1}$) for lithium batteries and high power density ($Wkg^{-1}$) of EDLC system [8–11]. Furthermore, few polymers depict several characteristics changes and some benefits in the potential application field in the presence of electrical stimulation; those are called electro-active polymers or EAPs. Those materials are showing response in size and shape by the variation of electrical stimulation. This material is categorised into two depending on the activation mechanism, i.e., ionic and electronic or field-activated [12, 13]. Generally, the electronic grade exhibits a higher energy density of mechanical energy; in comparison, ionic systems potentially employed ion diffusion or transportation consist of the electrolyte-electrode interface. Conductive polymers, ionic polymer gels, and composites of ionic polymer-metal are examples of some ionic EAPs. The electromechanical response of the membrane, showing huge strain at electrical stimulation, is very similar to the biological tissues. The Ionic polymers are introduced in fuel cells application in the 1960s; the features of EAP and the connexion of electromechanical ion-transportation of metal-ionic polymer composites (IPMC) were invented in 1992 by researchers in the United States and Japan [14, 15]. In general, IPMC comprises a membrane of polyelectrolyte, normally Flemion or Nafion covered on all sides with a conductive metal. After that, the counter-ions are neutralised, stabilising the electrical anionic charge of the covalently stable to the pillar of the membrane. The hydrated cations transportation between an IPMC polymer matrix is kept within the voltage applied, and corresponding electrostatic interactions enhance the bending. IPMCs are the working actuator that delineates lower impedance. Moreover, the water-based IPMCs generally dissociate with solvent content when it is subjected to 1.23 Volts higher. IPMCs are corroborated as the smartest prominence materials for larger bending deformation and lighter weight within significantly less applied voltages [16–18]. However, To fabricate chemically coated IPMC membrane, electrodes of metal ions gold, platinum, etc., are distributed through the hydrophilic sections of the polymer matrix and generally decrease the consistent metal atoms of zero-valence. Paddison et al. [19] researched the measurement of hydrated Nafion's permittivity in the range of broadband frequency. Their outcomes denote that the dielectric constant enhances with incrementing the water content, reduced with frequency increase.

## 2. Different manufacturing process of IPMC membrane

The important pillar for IPMC manufacture procedure is choosing the base ion altercation polymers; those are naturally fabricated from the organic polymers, comprising a permanent ionic group of covalently bonded. In the electrochemical industry, generally well-known ion exchange materials are used that rely on a copolymer of divinylbenzene and styrene. Moreover, the permanent ionic groups are shaped after the completion of polymerisation. The widely held ion altercation materials are alkene of perfluorinated: with lesser length side-chains concluded by ionic groups like Nafion, normally ammonium cations for anion exchange or carboxylate or sulfonate ($COO^{-}$ or $SO^{3-}$) for exchange of cations. In a perfluorinated compound (PFC), all hydrogen is supplanted by fluorine in the chain of carbon; however, the molecule holds one altered functional group or atom at least. The backbones of large polymer conclude their short side-chains and mechanical strength afford ionic groups that interrelate with water and the transportation of suitable ions. Furthermore, they might generate nano-channels of hydrophilic nature that are called cluster networks. Yu. et al. was fabricated the IPMC polymer matrix using the as follows method. However, in this process, the perfluorinated ion exchange membrane is taken upon that metal is deposited, such as platinum or gold. The IPMC matrix samplings were constructed with the twice depositing of platinum on Nafion–117 [20]. The initial stage is to coarsen the material surface; a) the emery paper is used to scratch the membrane surface for increasing the effective surface area; b) using an ultrasonic cleaner, the membrane is cleaned with preferable water; c) the membrane is dipped and boiled the in aqueous hydrochloric acid, i.e., HCl aqueous of 2 N concentration for 30 minutes to eradicate the ions and impurities in the membrane, d) finally after the rinsing with DI water, the entire membrane is merged in hot deionised (DI) water for up to 30 minutes to eliminate acid and swelling the polymer membrane matrix. The coarsened membrane is kept in deionised water. The subsequent procedure steps are to integrate the ion transportation or altercation capability with the help of a complex metal solution. The platinum amine complex ($[Pt(NH_3)_4]Cl_2$) aqueous solution is used. Furthermore, when the polymer membrane is submerged, a solution of ammonium hydroxide (5%,1 ml) is added that to neutralise. Store the membrane at room temperature in the solution for more than three hours. Following Yu et al., the third step is primary plating. In this procedure, the complex of platinum cations is decreased to a metallic state as nano-particles form with the help of a prominence-reducing agent. After the rinsed in 180 ml 40°C deionised water, sodium borohydride (2 ml of 5%) is mixed in each 30 min for seven times. Within this arrangement, the temperature is raised progressively to 60°C for 1.5 hours. A smooth Pt particles black layer is coated in the membrane surface. After that, the entire polymer matrix, i.e., the membrane, is again washed with deionised (DI) water and plunge in 0.1 N dilute hydrochloric acid for up to one hour. In the finishing procedure, the secondary layer of plating is completed. This methodology is envisioned to deposit Pt over top of the preliminary surface of Pt to diminish the resistance of the surface. Next, the supplementary quantity of Pt is overlaid with the following methods into the as grow Pt layer. Furthermore, hydroxylamine (6 ml) and 5% hydrochloride solution are added, and 20% of hydrazine solution (3 ml) is mixed in every single 30 minutes interval simultaneously. In this sequential fabrication route, the temperature is elevated to 60°C progressively for four hours; after that, the grey metallic texture will be coated on the membrane surface. After that, when no platinum ions are there in the solution of plating, the membrane is adequately washed with water and again placed in boiling in 0.1 N hydrochloric acid (dilute) to eliminate the cations of ammonium of the membrane.

## 2.1 Casting method and silica gel process

The economically available Nafion film thickness varies in the range of $50 \sim 180$ μm, though the efficiency of the IPMC differs by thickness, like the rigidity of bending, a field of electric, etc. However, to attain a denser or targeted depth of Nafion membrane irrespective of customised thickness, the entire process is illustrated below. Furthermore, this procedure needs a very controlled tuning of the method's parameters like the specific solvent concentration and temperature; therefore, it has reproductivity issues [21]. The thickness of Nafion may be attuned by the fine-tuning of volume dispersion of the Nafion. The overall fabrication procedure comprises four steps (a) stirring, (b) mixing and (c) thermal treatment, and (d) sonication. Thermal handling enhances the Nafion film's mechanical stiffness. Lastly, the Nafion film must be placed in the hydrogen peroxide solution in boiling state for 1 hour within temperature range 75°C -100°C and then the membrane heated for 1 hour in the deionised (DI) water. Furthermore, the process of casting has been testified to attain higher thickness Nafion IPMC actuators [22]. Moreover, Jung et al. was established a Nafion membrane of controlled pore size with porosity by silica sol–gel methods and etching by hydrofluoric acid [23]. Furthermore, they established an improved metal composite actuator consisting of ionic polymer with the help of Nafion membrane having porosity using the ion exchange procedure and electroless plating. After that, (i) the different Nafion surface possessing membranes were coarsened with sandpaper #1200 grading, and after that, the surface is substantially cleaned by ultrasonication cleaning method and chemically using HCl solution (2.5 N); (ii) for the exchange methods of Pt ion, the membranes were dipped in $[Pt(NH_3)_4]Cl_2.xH_2O$ solution (0.2 wt%), and then ultrasonicated for 8 h with 30 min interval; (iii) for the initial reduction methods, an aqueous solution about 180 ml having NH4OH (0.5 ml) solution was arranged, and then the membranes are absorbed in that solution; (iv) 1 wt% (0.5 ml) PVP solution and 2 mL of 5 wt% $NaBH_4$ solution is mixed in each 10 min interval; and (v) while the temperature of the solution is 600°C, the 20 mL of $NaBH_4$ solution (5 wt%) was mixed, and after that PVP solution (0.5 ml 1 wt%) was mixed at intervals of 20 min for four times. Finally, while this method was concluded, the membranes were dipped for 12 h in 0.1 N HCl solutions and then cleaned with deionised (DI) water. Thus, the membranes were accomplished by the Pt ion exchange methods of four cycles and the methods of initial reduction. Also, in the second-reduction methods, an aqueous solution of 280 mL is made that contains $[Pt(NH_3)_4]Cl_2.xH_2O$ (400 mg) was executed, and the membrane was dipped in that solution at 400C. Then 3 mL of NH2OH-HCl solution (5 wt%) and 6 mL of $NH_2NH_2.H_2O$ solution (20 wt%) were mixed with each 10 min. After that, the membrane was cleaned with DI water and 0.1 N HCl solutions. Lastly, in cation exchange, the membranes were dipped for 48 h in 1.5 N LiCl solutions (**Figure 1**).

## 2.2 Nafion-CNT composite based IPMC membrane

Ijeri et al. have prepared a carbon nanotube and Nafion polymer nanocomposite for combinational electron and proton transportation. The membranes depict the proficiency to permit a separate electron-proton transportation path inside the

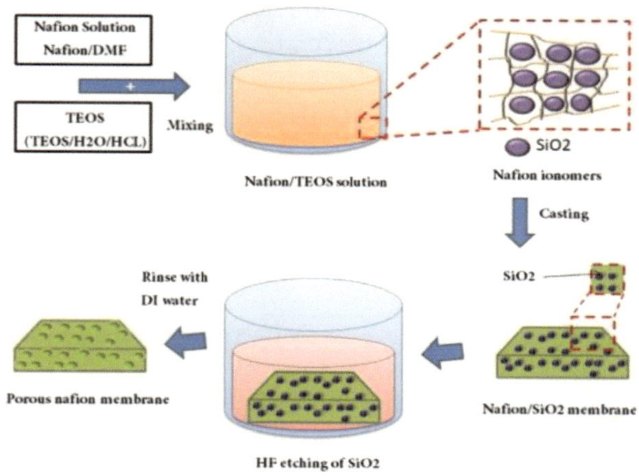


**Figure 1.**
*Schematic representation of the fabrication of Nafion membrane technology incorporated with porosity [23].*

membrane matrix. This feature unfastens new applications in particular membranes like advanced synthetic devices proficient in working sunlight to harvest hydrogen using water splitting. The manufacturing procedure is delineated as follows [24]. First, the needed amounts of MWCNTs and Nafion solution were weighed and put for 20 min in an ultrasonic cleaning bath with water. This provided a uniform dispersal of MWCNTs. The quantities were designed to provide MWCNT (0–5%) on a dry weight basis in nafion. In addition, isopropyl alcohol (1 ml) was mixed to assist in the proper dispersion and wetting of MWCNTs. Whenever the homogenous solution was decanted into petri dishes, it was kept on a furnace, maintaining that it is placed in an equal height levelled flat platform allowing uniform evaporation of solvent for 3 h within ambient conditions. Moreover, the entire setup was then replaced at 40°C and in the oven overnight. Furthermore, the membrane was then moistened with deionised (DI) water for one hour, separating the membrane from the entire setup substrate. Finally, self-supporting, flexible membranes were fabricated by just detaching them from the entire petri dish. Finally, the membrane was dried again and kept in an oven overnight.

### 2.3 Fabrication procedure by silver nanolayer in the membrane

Chung et al. introduce nanopowder of silver for IPMC fabrication membrane, which progresses the adhesion between polymer membrane matrix and metal electrode [25]. The normal methodologies of this process are the nanopowders of silver casting with Nafion polymer trailed by ornamentation and technologies of electroless plating of silver. The IPMC actuator of 5 mm (width) x 2 cm (length) x 0.23 mm (thickness) is taken for deformation investigation. The IPMC membrane corroborates a huge bending deformation with more than $90^0$ angle curvature bending at 3 V applied. For the Ag nanopowder casting, the in-between adhesion of the electrode of metal and the polymer matrix membrane delineates superior attributes; however, this will enhance the surface resistance up to 2 Ω/square. To solve these difficulties, a non-toxic electroless plating of silver is introduced to diminish the resistance of the surface. Moreover, the resistance of the surface is lessened to 0.12–0.15 Ω/square after this electroless plating method. Furthermore, in the procedure, the contact pad will incline to form $Ag_2O$ for the oxidation of Ag by the influence of $OH^-$ from the water that will increment more the resistance of the surface. The nickel electroforming in the contact pad will lessen the surface's resistance because it will preclude the $Ag_2O$ formation. This novel is emerging procedure has the eminence of the better bond between the polymer membrane and electrode, shorter process time, substantial driving force in the lower driving voltage (around 0.22gf at 3 V), low cost those making it potential for application of mems based device (**Figure 2**).

## 2.4 The process of hot-pressing

The hot-pressing method is used to fabricate numerous thin films of Nafion observe together, which improves the stiffness of bending, reproducibility, and force performance. The key benefits of this process are that it is simple, repeatable, and lucid to regulate the thickness [26]. Mainly the actuators of IPMC are fabricated by this method. Modify the films of Nafion with appropriate dimensions, clean by the acetone, and pile by the polyimide film in the mildew. The mould setup is then positioned between the 180°C preheated setup with presses for 20 minutes without

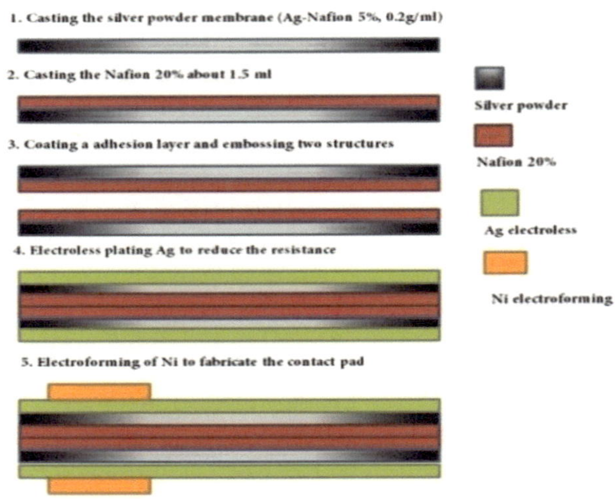


**Figure 2.**
*The schematic method flow of Ag nano-powders coated IPMC actuator [25].*

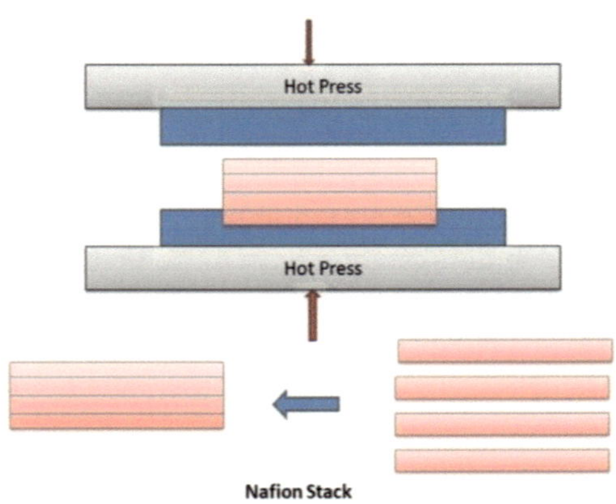


**Figure 3.**
*Fabrication by hot press and stacking [27].*

pressure, and at 180°C, then it is hard-pressed in the pressure of 50 MPa up to 10 min. After the films are cooled in ambient air to ambient temperature, the film is boiled for 1 hour at 70°C 3 wt% sulphuric acids for 1 hour at 70°C in 10 wt% hydrogen peroxide for 0.5 hours in DI water. Furthermore, dipped in platinum ammine complex ($[Pt(NH_3)_4Cl_2]$) aqueous solution for one to two days for platinum ions absorption in the piled film. Furthermore, after cleaning the membrane with deionised (DI) water, keeping it at 40°C water of 500 ml, stirring and mixed sodium borohydride (5 wt%, 5 ml) solution in 30 min every interval and also temperature steadily incremented up to 60°C (**Figure 3**).

## 3. Properties of IPMC

### 3.1 Actuation of IPMC membrane

When a membrane of Nafion-based IPMC polymer is induced to a slight amount of DC potential, the membrane experiences a deformation of rapid bending incline to the anode, trailed by a time-taking relaxation headed for cathode that is in the opposite direction. When both surfaces are connected after the movement of relaxation has clogged, the test sample exhibits a faster deformation of bending in the direction of cathode and, after that, gradually relaxes to its prior position to the anode direction.

### 3.2 Electrical and mechanical features of IPMC

It is established that the IPMC holds high modulus of elasticity and stiffness than the Nafion membrane, while both trails a similar pattern from the stress and strain analysis. This is because strain–stress performance is prevailed by the polymer than the powers of metal (as coated as electrode material). At the time, IPMC functions in a mode of bending; there is dissimilarity in mechanical features of the particles of metal (in the electrode) and network of polymer likely to influence one another. Moreover, fabricating a potential strain and stress plot of the IPMC membrane should be tested and cut as a cantilever configuration. The impedance analysis is delineated that IPMC works as an impartially capacitive material in low-frequency (>100 μF) and a resistive material like >50 Ω in higher frequency (**Figure 4**) [26].

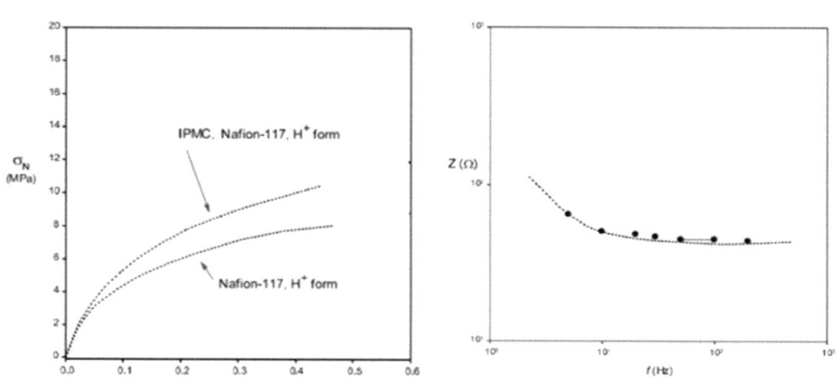


**Figure 4.**
*Impedance spectra of wet IPMC sample at fully hydrated state [27].*

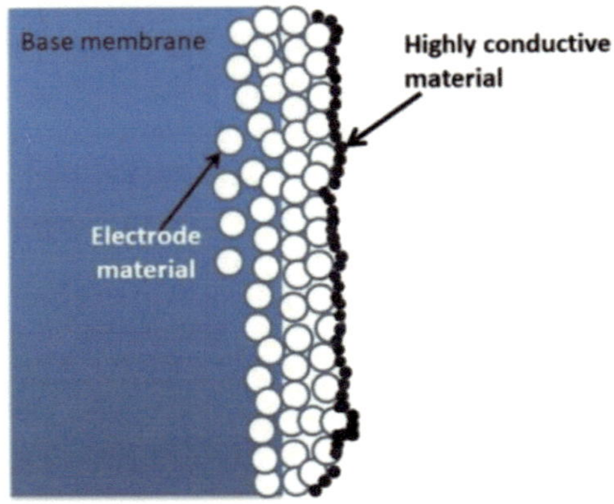


**Figure 5.**
*Schematic representation of the deposition process of higher conductive material [27].*

### 3.3 Effect of cations variation

Kim and Shahinpoor [28] investigated in detail to realise that IPMC polymer matrix features vary depending on cations variation. Each membrane was manufactured to absorb dissimilar counter-ions likely as $Li^+$, $Na^+$, $H^+$, $Ca^{2+}$, $K^+$, $Ba^{2+}$, $Mg^{2+}$ by sopping it in a particular solution of salt (1.5 N of LiCl, NaCl, HCl, $CaCl_2$, KCL, $BaCl_2$, $MgCl_2$), individually for 3 days at 30°C [6, 29, 30]. All the IPMC sample is monitored at null displacement holding different cations with respect to the $Na^+$ ions in IPMC. So it is clear that $Li^+$ – comprising IPMC is more outstanding than others, representing those hydration procedures compared to transportable cations that create a prominent character in actuation. Similarly, Flemion corroborates the same response (**Figure 5**) [31].

### 3.4 Lowering of IPMC electrode surface resistance

The resistance of the surface shows a crucial part in the electro-deformation of IPMC. Shahinpoor and Kim [32] abridge the resistance of active electrode surface effects on the prominence of artificial IPMC system is inclusive [33, 34]. It is observed under the scanning electron microscope (SEM), the surface cracks on the IPMC are easily visible, and pores are easily distinguishable. The relaxation and contraction are repeatedly shown when actuation bending of an IPMC generates more electrode surface cracks and depreciates the IPMC conductivity of the surface. If the IPMC electrode surface has a high resistance surface, the solvent molecules and cations membrane inside will transport to the outer side electrodes associated with the supply of power for the gradient encouraged in the field of electricity. The AC impedance analysis is an electrochemical method that helps to investigate IPMC artificial systems that elucidate the structure's equivalent circuit. The surface-electrode resistances (Rss), the polymer resistance (Rp), and impedance (Zw) for the charge transportation resistance nearby the surface electrode, with the double layer capacitance, relates in the interface of electrode-surface-electrolyte (Cd) and ionic polymer. In general, the surface electrode length (L) and the thickness of the surface electrode (t) play an essential role [27].

## 4. Energy storage study of IPMC polymer system

The electro-active IPMC material comprises a central layer possessed by a polymer with the upper and lower layer prepared of higher conductive electrodes.

The polymer that conquers the central layer has two main attributes: ion selectivity and permeability. These features are accomplished using polymers comprising organic ionic groups involved by covalent bonds to the polymer's backbone. Rely on the sign of the charge of the ionic groups existing in the polymer; it may be penetrable to cationic charge, i.e., positive and anionic charge, i.e., negative, or both. The polymers are commonly used with fixed ionic sulfonate groups for permeable positive ionic charges [35]. Furthermore, it is stated that one of the most standard groups of polymers worked in IPMC is "perfluorinated alkene," and specimens of these polymers are Nafion, Neoseptat, Flemiont or Selemiont, and Aciplext [35]. The electrodes founding the upper and lower layers of IPMCs are electrical conductors, investigated by very low electric resistivity. They are generally platinum (Pt) or silver (Ag) [36]. In the methods of fabrication of an IPMC, the electrodes deposition is presently made on the upper and lower active surfaces in three techniques: the route of "incorporation through reduction" (suffusing reduction methods) [29]; the physical casting procedure and the "direct mounting methods" (natural process assembly). The latter was established to elucidate the main difficulty of the two other manufacturing methods: the poor control in the deposition time of electrodes. The efficacy of the "integration through reduction" methodologies has created this the most worked, despite being the most time-taking and highly expensive [37]. IPMC membranes were fabricated using Nafion 117 polymer that is penetrable to positive ionic charges. The electrodes were deposited by high conducting metal like platinum or gold on the polymer using the "integration by reduction" procedures. Moreover, the disadvantage of holding poor control in the surface features of the electrodes typically reasons IPMC membranes to have different textures. The surface morphology and geometry of the electrodes of the IPMCs must be associated with their enactment, concomitant with the electrodes' electric resistance, and the consequential IPMC dielectric constant. Fundamentally, existing research work uses IPMC materials as actuation elements in electromechanical systems [30, 38–42]. A research was conducted on the capabilities of electro-active IPMC capacitors and the requirement of these possibilities on temperature [43–47]. Numerous IPMC membranes were observed with the same thickness but varying active surface areas. The membranes did not need any electrolyte that is why it is called "dry" strips with those capacitor elements. The polymer constituent of the membranes was acquired from a sheet of Nafion 117 and having a definite thickness. After cutting, a layer of chromium of 5 nm thickness was coated on the polymer surface, followed by a gold layer with a thickness of 100 nm on that same platform. Chromium was used to guarantee good adhesion between the gold and the polymeric matrix. The features of the voltage terminals of the composite material were chronicled within charge and discharge tests. All assays were done when charging the IPMC using a current power source, from which constant current charge was produced. Each IPMC was charged over fixed time using a fixed electric charging current in mA and then allowed to discharge, investigating the material property itself, i.e., how long they hold a charge. Moreover, the obtained results are calculated, and the exhibited capacitance value of specific and areal both have in the very higher range. The charging procedure is happened very fast up to the limit. The operating voltage was also perceived that the operating voltage is set up to that above which the IPMC got disrupted. Some of the used IPMC membranes were cut to test their storage capacitance scalability with their effective surface area. It was witnessed that a lessening of 20% of the surface area amounted to a reduction of 20% of its storage capacitance, such as a linear relationship [48]. Another research work verified a system for producing and storing electrical energy using IPMC devices and polymeric polyvinylidene difluoride (PVDF) piezoelectric devices to feed emitting organic diodes. For the electrical energy storage of IPMC, electro-active devices were worked, also for the production of electrical energy, PVDF piezoelectric elements were employed, both as-fabricated single and double layers. The feasibility of IPMCs as electric energy storage elements was measured. The experiments were executed concerning the duration of the electrical charge store in the IPMC and reached the voltage level when the charging mechanism occurred at a constant voltage and constant current. Mainly The Nafion polymer is used in the IPMCs. The chromium electrodes were worked with an intermediary layer between the gold and the polymeric matrix, and also, the electrical contact from the gold platform is easy to make up [49, 50]. The storage capability of the IPMC membranes was certified with the integration of lithium ions in their interior structure. This for the initial doping of the IPMCs matrix was executed by dipping the devices in LiCl solution. The fabricated morphology of IPMCs matrix and metal-doped IPMC membranes stored charge in the greatest amount within a particular time window compared to the other storage elements, especially electrolytic capacitors, which exhibited specific and areal capacitances, too high. This work allows us to determine the relatively short electric charge of this material (**Figure 6**).

### 4.1 Electric model representing IPMCs as electrical energy storage elements

Dependent on the IPMC element mode, a variation of its theoretical model had to be prepared. This section analyses the various aspects and effects of the entire force on the positive ionic charges when the IPMC stores electrical energy. Electrical forces: When a continual electric current is enforced on the IPMC matrix, this leads to the positive ions, which are primarily at rest within the negative ionic polymer charges, to transport in the direction of one electrode. However, the negative ionic costs in the polymer are permanent; they will not be transportable. The gap between the positive and negative ionic charges will then escalate to the genesis of an electric field inside the IPMC matrix; there is a potential difference in its terminals and electrical forces between the IPMC systems. Mass diffusion forces: Mass diffusion forces are important in the experimental stage. When an IPMC membrane is dipped in an electrolyte, the forces for the mass diffusion method contribute to the impregnation of the membrane with the component positive ionic charges of the electrolyte. At the time of positive ionic charges, concentration in the membrane equals outside positive ionic charges. The concentration on the outer surface is residual; a substantial concentration gradient can transfer charges into the polymer matrix. Nafion 117's polymer encompasses negative charges, making the IPMC membrane selective only to positive charges concerned by the electrical forces

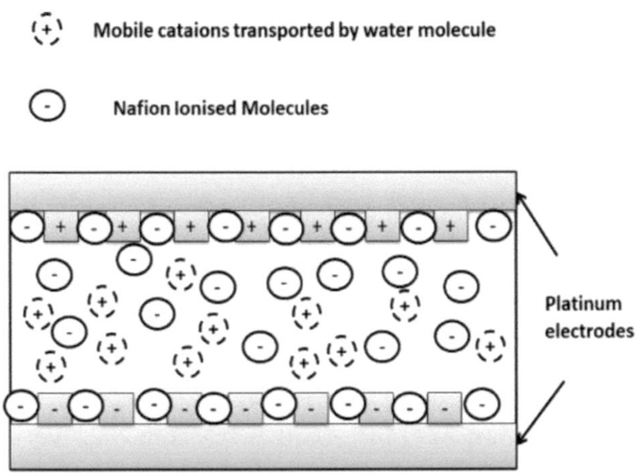


**Figure 6.**
*Illustration of ionic distribution in the IPMC based system [51].*

between them and the negative fixed charges. Mechanical forces: There are no mechanical pressures forced on the IPMC capacitor. The positive ionic charges current density: regarding the forces' characterisation, one arrives at Eq. (1) for the positive ionic charges current density within the IPMC.

$$J_{+} \approx -\left(\frac{RT}{K}\right)\nabla_{\rho c}^{+} - \left(\frac{q+}{kp}\right)\nabla_{pmec} \tag{1}$$

### 4.2 Discharging and charging of IPMCs at constant current

The discharge investigation under a resistive load intended to acquire the specific capacitance values $C_{eq}$ ($Fg^{-1}$) connected with a specific IPMC membrane, i.e., a capacitive element, and compute each discharge time $t_d$ and each used electrolyte. To execute a discharge test, it is essential to execute the electrical charging of each IPMC capacitive membrane. However, in every discharge, the test contained two divergent phases: discharging through a resistor and constant voltage charging. The obtained value was selected to have a much low value than the internal IPMC $R_{dif}$ resistance. This is expected to realise the synchronised use of two membranes of IPMC materials connected in parallel. The upper IPMC electrodes were associated with the positive terminal of the external DC voltage, while the lower electrodes were coupled to its negative terminal. There is a discharge resistance, i.e., $R_{ext}$ is present in between the terminals of the external circuit. Two of the most significant parameters for analysing a device for electrical energy storage are the rated voltage $V_n$ and charging time $t_s$. A series of IPMC charge tests are executed at a constant current to acquire values for those two parameters. The current source has the main benefits of being consistent for quite lower values of current in the mA order, and the voltage is controlled. The usage of the lower current value is for charging the IPMC is vindicated by these for being usable in low-power devices. This voltage source permits the constant current imposition to the IPMC capacitive membrane elements is crucial for the investigations of charging and discharging. The net gain in the current power source is better. The IPMC charging analyses were carried out at a constant current. This value was selected because the thinnest IPMC used in those tests reaches its nominal voltage of 1.5 V.

### 4.3 Duty cycle

The quantity of charge–discharge cycles provides the valuable life of a precise electrical energy storage device that can tolerate ago no longer being fit for operation. Those parameters are essential for indicating how long or for how many cycles the elements can be worked for a specific application. To conclude whether altered solute concentrations could enhance the useful life of an IPMC capacitive membrane, numerous consecutive charging and discharging investigation at constant current were executed. This harvests a square waveform of electric current conforming to charging and discharging cycles at fixed current. The waveform was asymmetric because a negative current in the discharge time was introduced, and the IPMC terminal voltage would reach negative values [51].

### 4.4 Holding time

Preferably, a device for electrical energy storing should deliver all of the electrical energy previously-stored irrespective of the time at which it was stored. The IPMC electromechanical model delineates that the electric charge in the IPMC system is related to the terminals voltage in Eq. (2). This designates that the electric charge stored in an IPMC is directly proportional to the voltage between its terminals, as in a capacitor.

$$Q = \frac{3\varepsilon bl}{5d} V \tag{2}$$

The experimental methods for assessing the conservation of electrical charge in IPMC capacitive elements had two dissimilar parts; whole charging of the IPMC following the analysis of its voltage at subsequent instants of time [52]. When the IPMC rated voltage is reached, the current source was turned off. The terminals voltages of the membrane were then calculated by varying times. The IPMC material depicts capacitive behaviour, and the value of its capacitance relies on the dielectric constant as in Eq. (3).

$$C_{eq} = \frac{3\varepsilon bl}{5d} \tag{3}$$

The specific capacitance $C_{eq\_e}$ (F $kg^{-1}$) of a given IPMC membrane is measured by Eq. (4), where r is the equivalent mass density of the IPMC membrane.

$$Ceq\_e = \frac{Ceq}{m} = \left(\frac{3\varepsilon bl}{5d}\right) \cdot \frac{1}{(\rho.b.l.d)} = \frac{3\varepsilon}{\rho d^2} \tag{4}$$

The dielectric constant value of the as-fabricated IPMC capacitive system is affected by the electrolyte used and the negative ionic charge density of existing connections to the polymeric morphology. Therefore, the connexion between the dielectric constants of two different IPMC membranes, if assembled in the same electrolyte, is only depicted by the ratio existing within the negative ionic charges density in the same element, which in turn will rely on the volume of the IPMC system (*b.l.d*) and the ionic charges density of the polymeric matrix used, k, stated by the following relation in Eq. (5).

$$\varepsilon \propto k.b.l.d \tag{5}$$

The significant time evolution of the voltage at resistance $R_{ext}$ throughout the IPMC discharge is given by Eq. (6), where U is the voltage at the preliminary instant, $R_{ext}$ is the electrical discharge resistance, and Ceq is the equivalent capacitance of the IPMC element.

$$\vartheta(t) = U.e^{-\frac{t}{R_{ext} \cdot Ceq}} \tag{6}$$

The primary technique contained the numerical approximation of the voltage curves acquired experimentally via Eq. (6). The methodologies of nonlinear least squares with the help of the confidence region algorithm were introduced to assess $C_{eq}$. The $R_{ext}$ had a value in the range kΩ. As the regression curve calculated was not a better approximation of the experimental data curve. Eq. (6) for the capacitance of IPMC equivalent presumes only a single capacitive effect in IPMC membranes. This is delineated for the transportation of positive ionic charges in between the polymer matrix, and it's accumulated along the whole surface of the electrode with a negative polarity (**Figure 7**).

Moreover, when the low-frequency is applied electrical signals like 0.1 Hz in this proportion of electro-active material, the positive ionic charges circulated the inner side of the IPMC matrix quickly gathered the nearby zone of the negative electrode, and consecutively the capacitive double layer is formed. The outcomes of

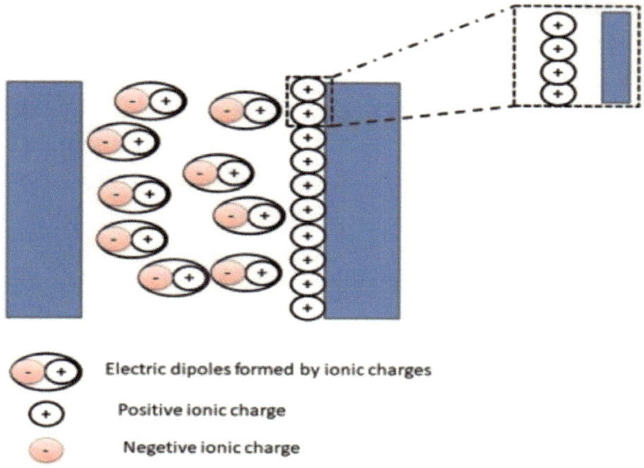


**Figure 7.**
*Schematic representation of the formation of the electric double layer in the matrix of polymer.*

experimental analysis, however, depicted the capacitive effect of higher potentiality. On the basis of the physical model formerly manifested for the IPMC materials, it was noted that the capacitive effect was only corroborated with the formation of electric dipoles in between the positive ionic charges and fixed ionic charges those are located in between the double layer region and the positive electrode. In conclusion, there are two distinct capacitive effects: one is for the genesis of the double layer, and another is associated with the produced electric dipoles. To interpret for the two capacitive effects in the time frame of the voltage, an amendment was made to Eq. (6), containing the sum of a second exponential component linked with a second time constant, as Eq. (7) shows [52].

$$\vartheta\,(\mathrm{t}) = \mathrm{a}.e^{\frac{-t}{\tau 1}} + b.e^{\frac{-t}{\tau 2}} \tag{7}$$

The model now reflects two-time constants: a short time constant $\tau\,1$ and a slow time constant $\tau\,2$. The sum of those parameters a and b is equal to the IPMC voltage at the initial stage of the discharging methods. The capacitance value is obtained related with each time constant from Eqs (8) and (9). One of that time constants is associated with the capacitive effect in the IPMC linked with the electric dipole arrangement and which zone in the polymer resembles the region between the positive ionic charges and the fixed ionic charges. On the contrary, the second constant is allied with the double layer.

$$\mathrm{C1} = \frac{\tau 1}{Rext} \tag{8}$$

$$\mathrm{C2} = \frac{\tau 2}{Rext} \tag{9}$$

Using the revised model, a substantial enhancement was achieved, allowing assessing the value of capacitances C1 and C2 and the time constants connected with each type of IPMC system and for each type of electrolyte. The fact that the maximum capacitance was accomplished when using electrolytes with lower solute concentrations can be elucidated by the encapsulation effect [53]. The encapsulation effect ascends from the circumstance that the maximum number of electric dipoles designed in the polymer matrix is attained for a given solute concentration of the electrolyte. A definite limited number of electric dipoles will relate to a maximum yield of the dielectric constant of the IPMC matrix. Since the capacitance concomitant with an IPMC is correlated to the dielectric constant using Eq. (9), it follows that the capacitance will incline to a maximum value at high solute concentrations. It should be recollected that $\tau 1$ is associated with the polymer region where the creation of electric dipoles in between positive ionic charges and fixed ionic charges happens and the region that will establish the characteristic times of charging and discharging of each IPMC system. However, an IPMC membrane having a higher thickness does not signify a greater time constant but directly proportional to the effective surface area. The difference in surface areas also delineates the alteration in results between the IPMC with the same thickness. Therefore, the second-largest surface area of the IPMC membrane in the research would be predictable to have higher yields than were acquired.

### 4.5 Discharge time

For an RC circuit, the load voltage during discharge of the capacitor over resistance is correctly given by Eq. (10).

$$\vartheta(\mathrm{t}) = \mathrm{U0}.e^{-\frac{t}{R.C}} \tag{10}$$

The discharging time constant $\tau$ is given by eqn

$$\tau = R.C \tag{11}$$

Replacing Eq. (11) into Eq. (9.20), one arises at the voltage at time t given by Eq. (12).

$$\vartheta(\mathrm{t}) = \frac{U0}{e} \approx 0.368Uo \tag{12}$$

The instantaneous at which the IPMC voltage reaches the value calculated by Eq. (12) is measured from all the experimental results. It was depicted that the majority of IPMC membranes offered results in the order of hundreds of seconds. Incrementing the values with enhancing electrolyte solute concentrations were also found. This system had the lowest surface area among the IPMC matrix analysed, thus having fewer electric dipoles along the electrodes of this matrix. The IPMC double-capacitance model undertakes two different capacitive effects; one is related with a fast time constant $\tau 1$ and with the other a slow time constant $\tau 2$. The significance of each time constants in the IPMC capacitor process is related to the frequency of the circuit in that it will be implanted. The capacitance C1 is leading in the case of high frequencies. Contrariwise, in low frequencies, the capacitance C2 will hold huge prominence in the operation of the IPMC capacitor. However, different functional features rely on the frequency operation of the circuit in which the IPMC is implanted; it is crucial to envisage its energy depending on the envisioned mode of operation. To calculate that percentage of the total stored energy can be free if high frequencies are used, the power degenerate in the resistive load in the time interval consistent to the first time constant—that is, from the initial time of the discharge until the time instant $\tau 1$—was calculated. Within these essays, two charging times importances had been taken care of analogous to two different time instants in the charging methods of the IPMC. The first time moment t corresponds to the instant at which the IPMC voltage value at its terminals is approximately 63% of the final value U, as shown in Eq. (13).

$$\vartheta(\tau) = \mathrm{U}\left(1 - \frac{1}{e}\right) \approx 0.632U \tag{13}$$

The second time instant is related to when the across IPMC terminals voltage equals 95% of its final value. These two moments were selected to depict the effect of the two-time constants of the system forecasted by the attuned electrical model. The best times were usually realised for higher solute concentrations, with the maximum having reached for IPMC using the higher electrolyte solute concentration. The charge time values are alike to the period of the cut-off frequency an IPMC element replies to mechanically. The charging time of an IPMC element is directly associated with its frequency response; meanwhile, this time interval resembles the time needed for much of the positive ionic charges stored inside the element is located on one of its electrodes, establishing dipole electric ionic in between positive and negative electric charges. Thus, one can authorise two charging time constants: a fast constant, which impacts the early charge stages, and a slow constant, with greater effect in the remaining moments. For cases of constant current charging, the model that expresses this fact is given by Eq. (14), where Umax is the IPMC voltage at the end of charging and the sum of the constants a and b is equal to 1.

$$\vartheta(\tau) = \mathrm{Umax}\left(1 - ae^{\frac{-t}{\tau 1}} - be^{\frac{-t}{\tau 2}}\right) \tag{14}$$

### 4.6 Nominal voltage and number of charge: discharge cycle

To assess an IPMC matrix system's electric power density over the maximum energy stored for each IPMC, the rated voltage must know first. It has been contemplated that the nominal voltage of an IPMC polymer matrix to the extreme potential variance can happen at its terminals deprived of the electrolysis of the solvent. Electrolysis of water is a physical circumstance investigated by the water decomposition into its basic elements, namely oxygen molecules and hydrogen ions. This incident has the effect of concentration lowering of solvent present in the IPMC, reducing the ionic mobility within the IPMC. Moreover, the H+ ions formation enhances the density of positive ionic charges, leading to a temporary increment in the ionic current density. If an adequately high electric field is extended, the polymer matrix electro-active IPMC membrane material interruption may still happen, that instigating permanent destruction to the material. The target is to inspect the association between the number of charge–discharge cycles and the solute concentration of a given IPMC polymer matrix. Different IPMCs were evaluated rely on the electric energy originally stored in certain IPMCs equated with the values at the end of the limited number of charge–discharge cycles of the IPMC. It is noted that these results were also equated with all solute concentrations considered. The methodologies introduced to relate the stored energy at the initial and end of a test relates to the square of the voltage at IPMC terminals. After a particular time, the outcomes for the decrease in stored electric energy were calculated as a solute concentration function. When the IPMC system is charged up to a definite maximum voltage, and after a few seconds, the entire system allows for discharge typically without external influence, and the total charge dissipation has occurred consecutively. The entire time taken for discharge is fully its charge storing capabilities. The primary rapid decrement in the voltage is due to the downfall of the double layer shaped by the positive ionic charges and the electric ones on the electrodes. The electrical energy stored in a capacitive system is given by Eq. (15).

$$\mathrm{E} = \frac{1}{2}.C.V^2 \tag{15}$$

In the time interval, the main incident was the reorganisation effect of the electric dipoles formed by ionic charges in the IPMC polymer matrix system.

As a result, the voltage at the IPMC lessened significantly. This voltage drop can be elucidated by evaporation of electrolyte, i.e., water in this case. However, the IPMC does not encapsulate; the evaporation of water will play a vital role in these circumstances [53]. As the solvent vaporises, this took away positive ionic charges, so the electric charge decreased within the IPMC system and the decreased number of electric dipoles. In some cases, it was observed that the membrane had dried, probably having some water in the matrix inside, which elucidates the presence of a residual voltage at its terminals. The energy density of stored electrical energy: It has been calculated that this type of capacitive element—the IPMC material possesses that is similar in nature to that of classic capacitors when used for energy storage. It is known, electrical energy is measured from the rated voltage and the capacitance. The power is proportional to the nominal voltage square directly, and, in turn, the rated voltage is proportional to the IPMC element thickness directly since the electrolysis of the solvent is catalysed by the electric field present within the IPMC system matrix. The energy density is exhibited by the ratio between the extreme energy the IPMC can store and its respective mass. It has also been seen that the response that incrementing the thickness of the IPMC does not impact energy density enhancement. This is a significant response that depicts that the price of a membrane of electro-active IPMC material becomes high prominently with the thickness. It is crucial to mention that the electrolyte a solution of salt and water for research purposes. If an electrolyte having a higher dielectric constant, like an electrolyte composed of lithium and propylene, had been introduced, it would be anticipated that the electric energy storage capacity of the IPMC materials matrix would be prominently enhanced. Lithium ions ($Li^{2+}$) developing the solute of the electrolyte has a lesser atomic radius than sodium ions ($Na^{+}$) and thus have higher ion transportation capabilities within a membrane of IPMC material, which accelerates the ion interaction between the sulfonate ions ($SO^{3-}$) and the positively charged ions attached to the polymer structure [54]. Moreover, this electrolyte's degree of evaporation is very low, which advances the preservation time of the electric charge of a capacitive IPMC material matrix.

## 5. Different applications of IPMC based energy harvesting and storage

### 5.1 Metal composites based ionic polymer on microcellular foamed Nafion

The energy harvesting experimentation of the IPMC composite membrane was lead in the 20 Hz frequency or less. When vibrations of a definite frequency window were allowed, the microcellular foamed samples were permitted, and a prominent proportion of energy was gathered from foamed samples compared to non-foamed samples. The outcome value was transformed by computing the modification data attained by the standardisation for the obtained current. In the band of frequency, the subsequent data was transformed into the value of root mean square (RMS), and also a 20 s time from the entire 30 min time approximately delineated the higher data was designated and connived. Two polymer membrane matrices were investigated in similar analysis environments, and the average value was accepted. For every analysis scenario, we connived a graphical output of current versus time. That was corroborated by that foamed IPMC samples logged a band of high frequency with the higher value. Moreover, in the band of low-frequency, it is depicted that the IPMC samples of non-foamed higher energy harvest compared to the foamed system. This analysis is accredited to the much critical non-foamed samples movements. In the band of frequency about 10 Hz, it was investigated that the foamed system deviate much higher swiftly than the

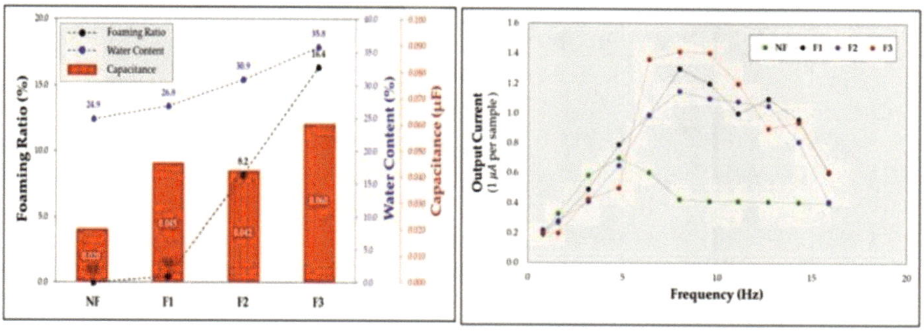


**Figure 8.**
*Water content, capacitance, and output current versus frequency variations according to the foaming ratio of IPMC specimen [55].*

specimen of non-foamed; also, the following data fluctuated up to roughly three times. The capacitance was calculated with the 0.001 μF resonance in the 120 Hz band after drying in a vacuum under all conditions of the specimen. Distinct outputs were gathered for every analysed system. As the ratio of foaming incremented, the retention capacity of cells' water was enhanced thereafter, along the capacitance was also incremented. It is presumed that the thickness increment in the membrane of polymer electrolyte and enhancement of water retentivity ameliorate the effective enactment of the capacitance. These are for the containing of water in the IPMC matrix and affecting the responses as a relaxation factor depicts the prominence outcome of the capacitance [55]. The obtained yields are delineated in **Figure 8**.

### 5.2 Ionic polymer-metal composite based water electrolysis for solar energy storage

The typical gas generation rate was calculated using the total volume of hydrogen gas composed divided by the duration of the 300 s. When the applied voltage increases, the gas generation rate enhances for all three IPMCs. Fascinatingly, the etched IPMC delineates a steeper increase compared to both the sanded and control IPMCs. Even though sanded and etched IPMC have higher whole gas production rates than the control IPMC, the sanded IPMC's gas generation rate was higher at a lower voltage, whereas the etched IPMC's gas generation rate was higher the higher voltage. To calculate the efficacy of the different fabrication process, the average gas production rate, the average voltage supplied, and the average current through the IPMC was computed to regulate the system's competence. From the graph, all three IPMC exhibits higher efficiency at lower voltage and their efficiency decrement at a variable rate as the voltage applied increases. Sanded IPMC has the highest efficiency and highest gas generation rate around 3 V, successfully producing that increasing the surface area through sanding will advance the performance of IPMC for energy harvesting. Although etching has a higher gas generation rate compared to the control, it must be noted that its energy efficiency values are often lower compared to both the sanded and control IPMC. On the other hand, at lower supply voltages, the electrolysers at higher temperatures were lagging in the generation of gas. However, the gas generation rate was higher at higher applied voltages, signifying a higher average slope in applied voltage vs. gas generation rate. The proficiency of the IPMC electrolyser generally decreases as the voltage applied increases. However, for higher temperatures (45°C, 55°C and 65°C), an increase in the productivity of the IPMC generator is seen as the voltage applied surges from 4 V to 5 V. When the temperature is low (45°C), it behaves similarly to IPMC at

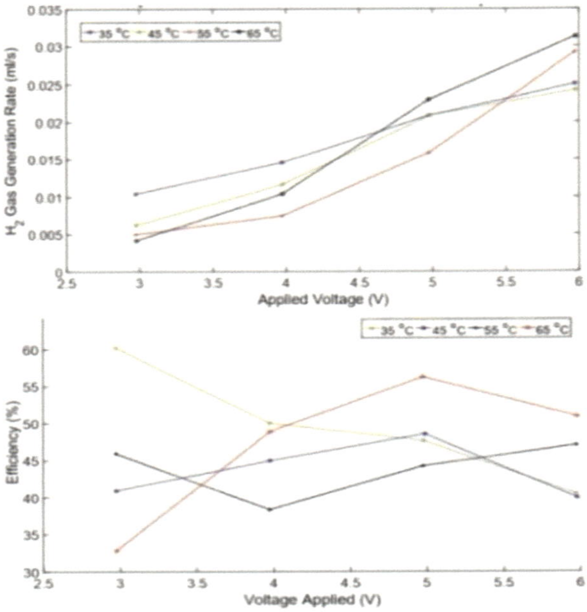


**Figure 9.**
*Comparison illustration of IPMC $H_2$ production rate and IPMC electrolyser's efficiency under temperatures variation [56].*

room temperature, however, when the temperature increases to 45°C, the electrolyser seems to hold a constant efficiency. As the temperature increases to 65° C, the efficiency seems to progress with increased voltage [56]. Comparison analysis illustrated in **Figure 9**.

### 5.3 Nafion-PVA composite membranes performance for direct methanol fuel cells

This research analysis has been targeted to investigate the permeability of methanol and fuel cell proficiency of Nafion-PVA composite membranes in the functionalities of the thickness in the range of 19 ~ 97 μm. Also, the composite polymer membranes were fabricated by the Nafion polymer that is placed in the nanofibres of (PVA) polyvinyl alcohol. The methanol infusion resistance of the Nafion/PVA composite membranes exhibits a linear deviation versus thickness. The variation between actual and apparent permeability leads to a calculated data of $4.0 \times 10–7\ cm^2s^{-1}$ for the proper or intrinsic permeability in the phase-in bulk of the membranes in the composite matrix. The integration of nanofibers of PVA creates a noticeable decrease of one order scale in the permeability of methanol associated with Nafion original membranes. The proficiency of DMFC of the electrode membrane assemblages fabricated from pristine and Nafion-PVA membranes was analysed at 95°C, 70°C and, 45°C in different concentrations of methanol, like as 3, 2 or 1 M. The membranes polymer composite in nanoscale with the thicknesses of 47 μm and 19 μm corroborated densities of power of 184 $mWcm^{-2}$ and 211 $mWcm^{-2}$ in the temperature of 95°C and the concentration of 2 M methanol. It is analogous to the finding for membranes by Nafion with an analogous thickness of the similar conditions, 204 $mW\ cm^{-2}$ and 210 $mW\ cm^{-2}$, correspondingly. However, a higher degree of utilisation of Nafion work as a material of proton-conductive in Nafion-PVA membranes is delineated for the lesser proportion of Nafion polymer in the membranes composite. Therefore, substantial reserves in the expended Nafion amount are significantly achievable. Moreover, the PVA nanofibers incorporation in the time of fabrication exhibited the polymer membranes with lower thickness with higher mechanical attributes; however, the modification of membranes of pristine Nafion become unviable below the 50 μm thickness. The innovative membranes of nanocomposite are fabricated using Nafion polymer amalgamated in between the functionalised nanofibers of polyvinyl alcohol (PVA) and investigated the attribution of DMFC performance and methanol permeability. Furthermore, methanol's primary permeability is detected from the attribution of original permeability inherent to the material of the membrane matrix. The composite polymer membranes delineated the permeability coefficient of methanol with a magnitude reduction of one order compared to the pristine membrane of Nafion for the barrier effect triggered with the nanofibers. Furthermore, the nanofiber phase is not impacted in the coefficient of electro-osmotic drag of methanol; in some definite circumstances, the low values were detected in membranes of Nafion-PVA. Fascinatingly, the coefficient of electro-osmotic drag of methanol was decremented corresponding to temperature as an alternative of the water responses conveyed to increment with temperature increase. Direct methanol fuel cell analyses at variational methanol concentration and temperature conditions delineated the extreme outcomes to be attained in 2 M solutions at 95°C. In these scenarios, the membranes of Nafion-PVA of 47 and 19 μm of thickness attained comparable engagements to Nafion membranes with equivalent thickness compared to the high protonic resistance detected at the membranes polymer composite. Introducing a phase of nanofiber in the Nafion polymer matrix and the thin membranes use leading to important reserves in the expended Nafion polymer amount can be proficiently managed to keep higher performances [57].

### 5.4 Ocean-based energy production system using the electrochemical alteration of wave energy with the help of ionic polymer-metal composites

In this research work, ionic polymer with metal composites (IPMCs) based on an energy harvesting platform stored the kinetic energy from the waves of the ocean and transformed it into electricity. However, the investigational analysis depicted that IPMC composite materials attribute several benefits, like durability and softness; they also counter speedily to wave parameters like wavelength, amplitude, and frequency. Moreover, the data analysis recorded for 296-day delineated that the gross power density engendered persisted stability around 245 $\mu W/m^2$. The decaying electrical performance of IPMC polymer matrix in a span of long term process is trivial. Generally, the modules rotation is nominal: the 18 modules' gross amplitude in the motions of rolling, pitching, and yawing is roughly 1 degree in the 0.48 Hz frequency.

Furthermore, the displacement in the z-direction is profoundly high compared to the incident waves' amplitude and different displacements directions. The investigational outcomes demonstrated the gross power density intensely oscillated at a particular time in a day more than 180 $\mu W/m^2$, also with the power density average of roughly 245 $\mu W/m^2$ and 292 $\mu W/m^2$ peak value. The IPMC polymer system's power density is mostly smaller than the conventional energy resources; furthermore, the IPMC system is economical and sturdy. The system power can be enhanced by accumulating the effective area of the IPMC membrane matrix. On the other hand, several explorations will emphasise evaluating buoy systems in fluctuating wave and current presences to analyse the configuration system stability in the ocean at the time of extreme weather or typhoon. Those studies will assist in augmenting the long-term performance and lessening the prices of maintenance at the installation time. Furthermore, the IPMC material bending plays a potential role to develop the proficiency of the energy harvesting structure [58].

## 5.5 Water and ion transportation in Perfluorosulfonated ionomer membranes for application of fuel cells

The volume fraction of water θ rises with lessening equivalent (EW) weight values of the IPMC membrane, irrespective of the membranes types and the counter cations. That point denotes the water present in the membranes matrix escalates with the concentration surges of the group of ion exchange. Afterwards, the water molecule presents in the groups of sulfonic acid and the counter cations within membranes to create the regions of the ionic cluster. On the other hand, anticipated that the membranes possess lesser EW values help to advance more numerous and much extensive ionic cluster zones in the membrane by providing high ionic conductivity. However, for the variation of cations species, the content of water order is $Na^+ < H^+ < Li^+$ for all the matrix of membranes. It designates that the content of water relies on the species of cations, and specifically, Li + cations can fetch the molecules of water for having the higher hydrophilic features. The ion-exchange group contributes to the formation regions of the ionic cluster. The total water molecules number in the membrane per ion-exchange group is $\lambda$ also enhances with lessening the equivalent weight value; also, the order of $\lambda$ for cations species variation is similar with the volume fraction of water $Na^+ < H^+ < Li^+$. It is advised that the number and size of regions of a cluster of ions in the matrix of membranes are pressurising with the value of the equivalent weight and the cations species variation. This signifies that when the equivalent weight diminishes, the counter cations and groups of sulfonic acid attract much water in the ionic ambience. The content of water is calculated by the stability in between the force of osmotic hydration in the ion and elastic force cluster persuaded with the fluorinated polymer deformation that is the leading chain. The activity of water can be equally presumed in the membrane matrix and the outer side of the solution of the membrane.31 If the number and size of the cluster region of ions enhance by escalating the membrane volume, the membranes density will be diminished for the water thickness (ca. at 25°C 1 g cm-3,) is lesser than the membranes itself (at 25°C around 2 g cm-3). The membrane densities reduce with reducing the equivalent weight value, particularly the Li-form membranes thicknesses that are lesser than the different cations-form membranes.

Furthermore, the membrane possesses a lesser equivalent weight value that enlarges the volume, and the cluster regions of ion are primarily developed mostly, specifically in the membrane of Li-form. The membrane possesses a lesser equivalent weight value that can create greater expanded cluster regions of ions despite the cations' species. The membranes' ionic conductivity is roughly correlated with the number and the size of cluster regions of ions in the matrix of membranes. Furthermore, the conductivity of all protonic membranes is significantly greater than the rest of the membranes of cationic form. It is usually attributed that the proton of the aqueous solution is elated with the help of the hopping process and can be transported faster compared to the rest of the cationic species transferred using the vehicle process. The probable aspects of finding out the conductivity are the mobility and concentration of carrier cations within the matrix of membranes that is calculated from the outcomes of the density of membrane and conductivity of ions. The sulfonic acid group concentration within the membranes, i.e. $C_{SO3-}$ is analysed and equivalent to the carrier cations attention. However, the carrier cations concentration for all membranes is impervious to the equivalent weight value of membranes. The carrier cations species mobility is enhanced with reducing the

EW value. Like this, the increment of carrier cations mobility becomes an initial factor in enhancing the membranes conductivity. Moreover, the perfluoro sulfonate ionomers are not membranes of the cross-linked polymer. The EW value decrease leads to increasing the cations species number and expanding the membrane volume for the swelling.

The motion of water molecules in the polymer membranes is subjective with electrostatic counteraction by the cationic species; the water molecules mobility within membranes are diminished when the stronger interaction is present, and the membrane's diffusion rate will be slower with the reduction of water molecules. Furthermore, the cations species and water molecules' mobility in the polymer membranes is prominently crucial to comprehend the ionic movement in the membranes. In this scenario, the coefficient of water transmission $tH_2O$ and the permeability of water Lp are significant parameters; they also provide such particulars about transportation. One carrier cation dragged that much water molecules denoted $tH_2O$ holds a positive charge when cation transports within membranes. If this value becomes higher, the molecules of water intermingle with the cation species to the membranes more sturdily.

On the other hand, *Lp* depicted that the rate of permeation of water molecules within membranes. The diffusion of water molecules arises more smoothly in the membranes when the value is higher. The *t*H2O value of the $Na^+$ and $Li^+$ form membranes depicted high values compared to the membranes of the H-form. The *Lp* Na-form and $Li^+$ membranes are lesser than the membranes of the H-form. The $Na^+$ and $Li^+$ cations have sturdier counteractions compared to protons in the membranes with water molecules and avoid the diffusion of water molecules.

Moreover, the yield results delineated of the H-form membranes is the transportation of proton controlled by the hopping process, and the molecules of water are dragged hardly together at the time of proton transportation within membranes. In this case, the water molecules easily diffuse compared to Na-and Li-form membranes. The variation in *Lp* and $tH_2O$ with respect to the EW value exhibits importance for water molecules with ion transport. The $tH_2O$ is not systematically varied in the proton membranes, with the increment of *Lp* with EW value decrement. The two distinct parameters, water permeability and water transference coefficient, have better associations with the membranes conductivity. It is corroborated that the water molecules move more easily, in which membrane depicted higher ionic conductivity. The water molecules mobility within membranes correlated with the conductivity and is motivated by the cation species and the equivalent weight value. To envisage the interrelation between the cation species and the water mobility in elaborate, the analysis of the coefficients of self-diffusion $DLi^+$ and $DH_2O$ was calculated by Li, and 1H PGSE-NMR are investigated. The values of $DH_2O$ are about $10^{-10}$ $m^2s^{-1}$ in the membranes. The acquired value for the proton Nafion 117 is very similar as investigated by Zawodzinski et al. in the temperature of 30°C. It is identified that the $DH_2O$ value of pure $H_2O$ at 30°C is $2.55 \times 10^{-9}$ $m^2s^{-1}$. So the molecules of water mobility in the membranes are decreased. The $DH_2O$ of the membranes enhances with reducing the EW value. The H-form membranes depict high values compared to the Na- and Li-form membranes. Those obtained outputs indicate that the water molecules transportation within membranes is decreased with the channel morphology of the cluster of ion zones and the interactivity by the species of a cation. Moreover, the molecules of water penetrate more rapidly in the Nafion membranes compared to the flemion membranes in the almost equivalent EW value. Also, it is fascinating that the polymer matrix membrane morphology impacts the membranes' conductivity. As a carrier ion, the movement of $Li^+$ is much connected to the water molecules transports in the membrane matrix. This is illustrated in **Figure 10**, here $DLi^+$ is illustrated versus $DH_2O$ of the similar

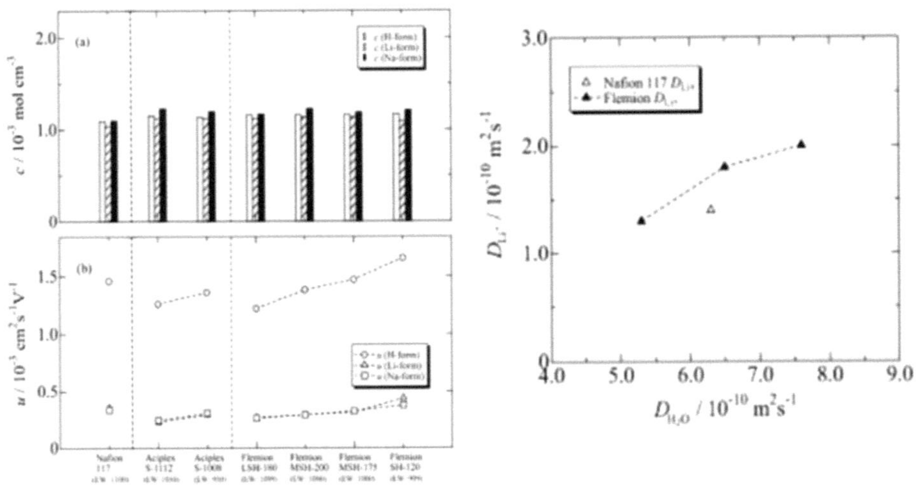


**Figure 10.**
*Mobility* u and *concentration* c *of carrier cations species in dissimilar membranes at the wholly hydrated state and the illustration of coefficient of self-diffusion of* $Li^+$ *cation* $D_{Li^+}$ *within Li-form membranes versus water molecule* $D_{H_2O}$ *[59].*

membranes. Generally, the diffusion phenomenon (small-range) is quicker than diffusion of longer-range is elucidated as inconsistent diffusion.

Moreover, the time-dependent $DH_2O$ illustrates that the molecules of water disperse in the heterogeneous region in the membranes. The areas of an ionic cluster are not the open area for the $H_2O$ diffusion, and the interaction of water molecules with $SO^{3-}$ and $Li^+$ shows dissimilar exchange rates. All of the obtained coefficients of self-diffusion analyses for $Li^+$ cation and water molecule by the PGSE NMR procedure are correlated well with other investigations of ionic conductivity, water content, water permeability, and water transference coefficient. Moreover, it is becoming a potential tool to comprehend water and ion-molecule transportation performance within the membranes [58].

## 6. Conclusion

This chapter represented the energy storage application with results and the practical characteristics of IPMC polymer materials in the analogous field. This composite polymer and polymer matrix have been researched in detail in the potential application of flexible energy storage of electrical energy. The attributes create the IPMC polymer materials prominent and potential elements for substituting the conventional capacitors in different real-time applications. These whole chapter goals delineate the summary of IPMC polymer material on energy storage and its corresponding working principles. There is plenty of fabrication methods comprising the primary process of compositing membrane and electrode of membrane surface discussed in detail. The coefficient of electro-osmotic drag of methanol is varied with temperature directly proportional. Direct methanol fuel cells analysed at different temperatures and methanol concentration delineated the highest performances in 2 M solutions at 95°C. The Nafion-PVA membranes of 47 μm and 19 μm thickness depicted proficient response, though the protonic resistance is high of composite membranes. The Nafion film thickness is enhanced the electrical harvesting response considering the more water retention in its internal pores. The foaming ratio escalation modifies the structure internally of electrolyte membrane and enhances the harvesting phenomena, and that corroborated to increase storage of electric charge and modification of capacitance, with augmentation of water. The content of water in the membranes depicted the propensity of the increment of ionic cluster areas number with size though equivalent weight value decreased and the Li + form membranes fashioned the biggest regions of the ionic cluster. Also, the coefficients of water transfer and permeability denoted that in the Na- and Li-form membranes, the interaction of water molecules much sturdily with cations compared to proton when it can transport within membranes, and the molecule of water diffusion is decreased. The research analysis illustrated that surface area increment of IPMC by the etching and surface sanding enhances the gas generation rate of the IPMC electrolyser. The temperature increment of water effect on the IPMC electrolyser and the analysis data depicts the sharp enhancement of gas generation rate with the voltage applied to the generator ameliorates. Generally, the module's rotation is very less along with the gross amplitude of the 18 modules in the yaw, rolling and pitching motions is roughly 1 degree at 0.48 Hz. Moreover, the displacement in the z-direction is substantially high compared to the incident wave's amplitude and displacements of other directions. The experimental analysis delineated that gross power density at a particular time of a day sturdily fluctuated above 180 $\mu W/m^2$, along with power density of roughly 245 $\mu W/m^2$ with 292 $\mu W/m^2$ peak value. This polymer matrix system and its prominence application in electrical energy storage are yet to investigate more, and there are lots of parts that have to be explored in upcoming times. This chapter will also be beneficial in comprehending the IPMC polymeric matrix system to plenty of interested researchers in electro-active polymers. To be precise, the application of IPMC materials in electrical energy storage acts as guidance.

## Acknowledgements

The authors wish to acknowledge the DST, Govt. of India, for financial support during the work.

## Author details

Monojit Mondal, Arkaprava Datta and Tarun Kanti Bhattacharyya*
School of Nanoscience and Technology, IIT Kharagpur, Kharagpur, W.B., India

*Address all correspondence to: tkb@ece.iitkgp.ac.in

## References

[1] Bandopadhya, Dibakar, and James Njuguna. "Estimation of bending resistance of ionic polymer metal composite (IPMC) actuator following variable parameters pseudo-rigid body model." *Materials Letters* 63, no. 9-10 (2009): 745-747.

[2] Brunetto, Paola, Luigi Fortuna, Salvatore Graziani, and Salvatore Strazzeri. "A model of ionic polymer–metal composite actuators in underwater operations." *Smart materials and Structures* 17, no. 2 (2008): 025029.

[3] Santos, J., B. Lopes, and PJ Costa Branco. "Ionic polymer–metal composite material as a diaphragm for micropump devices." *Sensors and Actuators A: Physical* 161, no. 1-2 (2010): 225-233.

[4] Tadokoro, Satoshi, Shinji Yamagami, Toshi Takamori, and Keisuke Oguro. "An actuator model of ICPF for robotic applications on the basis of physicochemical hypotheses." In *Proceedings 2000 ICRA. Millennium Conference. IEEE International Conference on Robotics and Automation. Symposia Proceedings (Cat. No. 00CH37065)*, vol. 2, pp. 1340-1346. IEEE, 2000.

[5] Wang, Xuan-Lun, Il-Kwon Oh, Jun Lu, Jinhun Ju, and Sunwoo Lee. "Biomimetic electro-active polymer based on sulfonated poly (styrene-b-ethylene-co-butylene-b-styrene)." *Materials Letters* 61, no. 29 (2007): 5117-5120.

[6] Keawboonchuay, C., and T. G. Engel. "Maximum power generation in a piezoelectric pulse generator." *IEEE transactions on plasma science* 31, no. 1 (2003): 123-128.

[7] Xu, Chao-Nan, Morito Akiyama, Kazuhiro Nonaka, and Tadahiko Watanabe. "Electrical power generation characteristics of PZT piezoelectric ceramics." *IEEE transactions on ultrasonics, ferroelectrics, and frequency control* 45, no. 4 (1998): 1065-1070.

[8] Mondal, Monojit, Dipak Kumar Goswami, and Tarun Kanti Bhattacharyya. "Lignocellulose based Bio-waste Materials derived Activated Porous Carbon as Superior Electrode Materials for High-Performance Supercapacitor." *Journal of Energy Storage* 34 (2021): 102229.

[9] Mondal, Monojit, Dipak Kumar Goswami, and Tarun Kanti Bhattacharyya. "Solvent dependent fabrication of Manganese Vanadium Oxide as cathode material for high performing supercapacitor." In *2020 4th International Conference on Electronics, Materials Engineering & Nano-Technology (IEMENTech)*, pp. 1-6. IEEE, 2020.

[10] Mondal, Monojit, Dipak Kumar Goswami, and Tarun Kanti Bhattacharyya. "Microwave synthesized manganese vanadium oxide: High performing electrode material for energy storage." *Materials Today: Proceedings* (2021).

[11] Mondal, M., B. Das, P. Howli, N. S. Das, and K. K. Chattopadhyay. "Porosity-tuned NiO nanoflakes: Effect of calcination temperature for high performing supercapacitor application." *Journal of Electroanalytical Chemistry* 813 (2018): 116-126.

[12] Bar-Cohen, Yoseph, and Qiming Zhang. "Electro-active polymer actuators and sensors." *MRS bulletin* 33, no. 3 (2008): 173-181.

[13] Bar-Cohen, Yoseph. "Electro-active Polymer (EAP) Actuators as Artificial Muscles: Reality." *Potential, and Challenges (SPIE-The International Society for Optical Engineering, Bellingham, Washington, 2001)* (2004).

[14] Oguro, K. "Bending of an ion-conducting polymer film-electrode composite by an electric stimulus at low voltage." *J. Micromachine Society* 5 (1992): 27-30.

[15] Sadeghipour, K., R. Salomon, and S. Neogi. "Development of a novel electrochemically active membrane and'smart'material based vibration sensor/damper." *Smart Materials and Structures* 1, no. 2 (1992): 172.

[16] Grodzinsky, Alan Jay. "Electromechanics of deformable polyelectrolyte membranes." PhD diss., Massachusetts Institute of Technology, 1974.

[17] Kim, Seong Jun, In Taek Lee, and Yong Hyup Kim. "Performance enhancement of IPMC actuator by plasma surface treatment." *Smart materials and structures* 16, no. 1 (2007): N6.

[18] Nemat-Nasser, Sia, and Jiang Yu Li. "Electromechanical response of ionic polymer-metal composites." *Journal of Applied Physics* 87, no. 7 (2000): 3321-3331.

[19] Peng, Jing, and Thomas A. Zawodzinski. "Describing ion exchange membrane-electrolyte interactions for high electrolyte concentrations used in electrochemical reactors." Journal of Membrane Science 593 (2020): 117340.

[20] Yu, Min., Shen, Hui., Dai, Zhen-dong., "Manufacture and Performance of Ionic Polymer-Metal Composites," Journal of Bionic Engineering, Vol. 4, No. 3, pp. 143−149, 2007

[21] Lee, S. J., Han, M. J., Kim, S. J., Jho, J. Y., Lee, H. Y. and Kim, Y. H., "A new fabrication method for IPMC actuators and application to artificial fingers," Smart Mater. Struct., Vol.15, No. 5, pp. 1217-1224, 2006.

[22] Kim, B. K., Kim, B. M., Ryu, J. W., Oh, I.-H., Lee, S.-K., Cha, S.-E. and Pak, J. H., "Analysis of mechanical characteristics of the ionic polymer metal composite (IPMC) actuator using cast ion-exchange film," Proc. of SPIE, Vol. 5051, pp. 486-495, 2003.

[23] Jung SY, Ko SY, Park J-O, "Park S. Enhanced ionic polymer–metal composite actuator with pore size–controlled porous Nafion membrane using silica sol–gel process," Journal of Intelligent Material Systems and Structures, Vol. 28(11), pp. 1514-1523, 2017.

[24] Ijeri, Vijaykumar., Cappelletto, Lucandrea., Bianco, Stefano., Tortello, Mauro., Spinelli, Paolo., Elena, Tresso., "Nafion and carbon nanotube nanocomposites for mixed proton and electron conduction," Journal of Membrane Science - J MEMBRANE SCI, Vol. 363, pp. 265-270, 2010

[25] Chung, C. K., Fung, P. K., Hong, Y. Z., Ju, M. S., Lin, C. C. K. and Wu, T. C., "A novel fabrication of ionic polymer-metal composites (IPMC) actuator with silver nanopowders," Sens. Actuators B: Chemical, Vol. 117, No. 2, pp. 367-375, 2006.

[26] Shahinpoor, M., "Ionic polymer-conductor composites as biomimetic sensors, robotic actuators and artificial muscles: a review," Electrochim. Acta, Vol. 48, No. 14-16, pp. 2343-2353, 2003.

[27] Bhandari, Binayak, Gil-Yong Lee, and Sung-Hoon Ahn. "A review on IPMC material as actuators and sensors: fabrications, characteristics and applications." *International journal of precision engineering and manufacturing* 13, no. 1 (2012): 141-163

[28] Kim, Kwang J., and Mohsen Shahinpoor. "Ionic polymer–metal composites: II. Manufacturing techniques." *Smart materials and structures* 12, no. 1 (2003): 65.

[29] Millet, P., M. Pineri, and R. Durand. "New solid polymer electrolyte

composites for water electrolysis." *Journal of Applied Electrochemistry* 19, no. 2 (1989): 162-166.

[30] Brufau-Penella, J., M. Puig-Vidal, P. Giannone, S. Graziani, and S. Strazzeri. "Characterization of the harvesting capabilities of an ionic polymer metal composite device." *Smart materials and structures* 17, no. 1 (2007): 015009.

[31] Onishi, Kazuo, Shingo Sewa, Kinji Asaka, Naoko Fujiwara, and Keisuke Oguro. "Bending response of polymer electrolyte actuator." In *Smart Structures and Materials 2000: Electroactive Polymer Actuators and Devices (EAPAD)*, vol. 3987, pp. 121-128. International Society for Optics and Photonics, 2000.

[32] Shahinpoor, Mohsen, and Kwang J. Kim. "The effect of surface-electrode resistance on the performance of ionic polymer-metal composite (IPMC) artificial muscles." *Smart Materials and Structures* 9, no. 4 (2000): 543.

[33] Shahinpoor, Mohsen. "Ionic polymer–conductor composites as biomimetic sensors, robotic actuators and artificial muscles—a review." *Electrochimica Acta* 48, no. 14-16 (2003): 2343-2353.

[34] De Gennes, P. G., Ko Okumura, M. Shahinpoor, and Kwang J. Kim. "Mechanoelectric effects in ionic gels." *EPL (Europhysics Letters)* 50, no. 4 (2000): 513.

[35] Kim, Kwang J., and Mohsen Shahinpoor. "Ionic polymer–metal composites: II. Manufacturing techniques." *Smart materials and structures* 12, no. 1 (2003): 65.

[36] Anand, S. V., K. Arvind, P. Bharath, and D. Roy Mahapatra. "Energy harvesting using ionic electro-active polymer thin films with Ag-based electrodes." *Smart Materials and Structures* 19, no. 4 (2010): 045026.

[37] Griffiths, David John. "Development of ionic polymer metallic composites as Sensors." PhD diss., Virginia Tech, 2008.

[38] Shahinpoor, Mohsen, and Kwang J. Kim. "Ionic polymer–metal composites: IV. Industrial and medical applications." *Smart materials and structures* 14, no. 1 (2004): 197.

[39] Peng, Wuxian, Yajing Zhang, Jinhai Gao, Yiming Wang, Yang Chen, and Yiran Zhou. "Fabrication and performance of ionic polymer-metal composites for biomimetic applications." *Sensors and Actuators A: Physical* 299 (2019): 111613.

[40] Bonomo, Claudia, Luigi Fortuna, Pietro Giannone, and Salvatore Graziani. "A sensor-actuator integrated system based on IPMCs [ionic polymer metal composites]." In *SENSORS, 2004 IEEE*, pp. 489-492. IEEE, 2004.

[41] Martin, Benjamin Ryan. "Energy harvesting applications of ionic polymers." PhD diss., Virginia Tech, 2005.

[42] Pandolfo, Anthony G., and Anthony F. Hollenkamp. "Carbon properties and their role in supercapacitors." *Journal of power sources* 157, no. 1 (2006): 11-27.

[43] Yeo, R. S., J. McBreen, G. Kissel, F. Kulesa, and S. Srinivasan. "Perfluorosulphonic acid (Nafion) membrane as a separator for an advanced alkaline water electrolyzer." *Journal of Applied Electrochemistry* 10, no. 6 (1980): 741-747.

[44] Conte, Mario. "Supercapacitors technical requirements for new applications." *Fuel cells* 10, no. 5 (2010): 806-818.

[45] Kaempgen, Martti, Candace K. Chan, J. Ma, Yi Cui, and George Gruner. "Printable thin film supercapacitors using single-walled carbon nanotubes." *Nano letters* 9, no. 5 (2009): 1872-1876.

[46] Kularatna, Nihal, Jayathu Fernando, Amit Pandey, and Sisira James. "Surge capability testing of supercapacitor families using a lightning surge simulator." *IEEE transactions on Industrial Electronics* 58, no. 10 (2011): 4942-4949.

[47] Landrock, Clinton K., and Bozena Kaminska. "Ionomer composite thin film capacitors." *IEEE Transactions on Components, Packaging and Manufacturing Technology* 1, no. 9 (2011): 1305-1310.

[48] Chuo, Yindar, Clint Landrock, Badr Omrane, Jeydmer Aristizabal, Jasbir N. Patel, Marcin Marzencki, and Bozena Kaminska. "Towards self-powering touch/flex-sensitive OLED systems." *IEEE Sensors Journal* 11, no. 11 (2011): 2771-2779.

[49] Branco, PJ Costa, and J. A. Dente. "Derivation of a continuum model and its electric equivalent-circuit representation for ionic polymer–metal composite (IPMC) electromechanics." *Smart Materials and Structures* 15, no. 2 (2006): 378.

[50] Invernizzi, Costante Mario. "Prospects of mixtures as working fluids in real-gas Brayton cycles." *Energies* 10, no. 10 (2017): 1649.

[51] Shahinpoor, Mohsen, ed. *Ionic Polymer Metal Composites (IPMCs): Smart Multi-Functional Materials and Artificial Muscles, Volume 2*. Royal Society of Chemistry, 2015.

[52] Robinson, Walter Junkin. "Charge control of ionic polymers." PhD diss., Virginia Tech, 2005.

[53] Kokya, Bahman Ahmadzadeh, and Taher Ahmadzadeh Kokya. "Proposing a formula for evaporation measurement from salt water resources." *Hydrological Processes: An International Journal* 22, no. 12 (2008): 2005-2012.

[54] Santos, J., B. Lopes, and PJ Costa Branco. "Ionic polymer–metal composite material as a diaphragm for micropump devices." *Sensors and Actuators A: Physical* 161, no. 1-2 (2010): 225-233

[55] Kweon, Byung Chul, Joo Seong Sohn, Youngjae Ryu, and Sung Woon Cha. "Energy Harvesting of Ionic Polymer-Metal Composites Based on Microcellular Foamed Nafion in Aqueous Environment." In Actuators, vol. 9, no. 3, p. 71. Multidisciplinary Digital Publishing Institute, 2020

[56] Keow, Alicia, and Zheng Chen. "A study of water electrolysis using ionic polymer-metal composite for solar energy storage." In *Smart Materials and Nondestructive Evaluation for Energy Systems 2017*, vol. 10171, p. 1017104. International Society for Optics and Photonics, 2017

[57] Mollá, Sergio, and Vicente Compañ. "Performance of composite Nafion/PVA membranes for direct methanol fuel cells." *Journal of Power Sources* 196, no. 5 (2011): 2699-2708.

[58] Vinh, Nguyen Duy, and Hyung-Man Kim. "Ocean-based electricity generating system utilizing the electrochemical conversion of wave energy by ionic polymer-metal composites." Electrochemistry Communications 75 (2017): 64-68.

[59] Saito, Morihiro, Naoko Arimura, Kikuko Hayamizu, and Tatsuhiro Okada. "Mechanisms of ion and water transport in perfluorosulfonated ionomer membranes for fuel cells." The Journal of Physical Chemistry B 108, no. 41 (2004): 16064-16070.

Chapter 7

# Structural, Optical, and Electrical Studies of PAN-Based Gel Polymer Electrolytes for Solid-State Battery Applications

*Vijaya Kumar Kambila*

## Abstract

Gel polymer electrolyte films (GPEs) based on polyacrylonitrile (PAN) complexed with NaF salt and an $Al_2O_3$ nanofiller were prepared via solution cast method. Structural studies were performed to investigate the order of conductivity under the influence of salt and nanofillers. The prepared films were characterized using energy dispersive x-ray spectrometry (EDS) to determine the chemical composition in wt%. EDS studies reveal that PAN–NaF with $Al_2O_3$ ceramic filler decreases the degree of crystallinity with increasing concentration of the nanofiller. The UV–Vis spectrum was recorded by a Hewlett-Packard HP8452A diode array spectrometer. The structural effect of salt and nanoparticles on the conductivity was also confirmed by UV–Vis spectroscopy. The mechanical properties of the prepared polymer electrolytes were determined using a Universal Tensile Machine (Instron Model 5565, Canada) with a constant crosshead speed of 10 mm/min. The addition of nanoparticles increased both the modulus and the strength of the polymer nanocomposites. Both the tensile strength and Young's modulus increased with increasing functionalized nanoparticle loading. The change in transition temperature caused by the incorporation of the $Al_2O_3$ nanofiller and plasticizer into the PAN+NaF complex was studied by differential scanning calorimetry (DSC) analysis. Additionally, DSC thermograms were recorded to measure the glass transition temperature and melting temperature of PAN-based electrolytes using a Mettler instrument. Conductivity studies were carried out for all the prepared polymer electrolytes to understand the conduction mechanism. The role of the ceramic phase is to reduce the melting temperature, which is ascertained from DSC. The sample containing PAN:NaF (70:30) exhibits the highest conductivity of $1.82 \times 10^{-4}$ S $cm^{-1}$ at room temperature (303 K) and $2.96 \times 10^{-3}$ S $cm^{-1}$ at 378 K. The polymer electrolytes considered in the present study exhibited an Arrhenius type of conduction. The polymer electrolyte containing 3 wt% $Al_2O_3$ nanofiller showed an ionic conductivity of $5.96 \times 10^{-3}$ S $cm^{-1}$. To determine transfer numbers, Wagner's polarization method can be used. From these studies, it is observed that the conduction mechanism is predominantly due to ions. Using this (PAN–NaF– $Al_2O_3$) (70:30:3) electrolyte, a solid-state electrochemical cell was fabricated, and its discharge profiles were studied under a constant load of 100 kΩ. Finally, several cell profiles associated with this cell were evaluated and reported.

**Keywords:** gel polymer electrolyte, solution casting technique, EDS, UV–vis method, tensile strength, solid-state battery discharge characteristics

## 1. Introduction

For the last three decades, ionic solid conducting polymer electrolytes have been used as active potential components in novel battery technology such as electrodes and electrolytes. They have excellent properties including light weight, appreciable mechanical strength, excellent plasticity, flexible processing, and easy fabrication in solid-state battery technology. The main advantages of solid polymer electrolytes are their chemical and physical stability, comparable performance to that of thin films (approx. 1 micrometer), and majority conduction of ions rather than electrons. However, the main drawback of solid electrolytes is interface stress due to electrode charging and discharging. Polymer electrolytes are used for batteries due to their ease of forming thin films with a large internal area; reduced resistance, which increases current density; stable and compatible contact with electrodes; and stability under ambient conditions such as temperature, pressure, and atmosphere to facilitate production on a mass scale.

Among polymer electrolytes, gel polymer electrolytes (GPEs) are a very important class of materials and have been used in electrochemical batteries such as fuel cells and electronic display devices [1–4]. Compared to liquid and solid polymer electrolytes, GPEs are found to be very advantageous. In general, in these types of GPEs, salt provides ions for conduction and solvents provide a medium for this conduction. Plasticizers also help to decrease the glass transition temperature. Sodium may be considered an alternative to lithium as a negative electrode due to its low cost and natural abundance. The softness of this metal enables better electrical property contact with other components in the battery. The addition of nanofillers such as aluminum oxide ($Al_2O_3$) to this type of salt-based film rapidly enhances the electrical properties. The present work reports on a polymer electrolyte (PAN+NaF + $Al_2O_3$) and is concerned with solid-state electrochemical cells that are based on (PAN+NaF) electrolyte films. Several experimental techniques, such as structural, optical, electrical, tensile strength, differential scanning calorimetry (DSC), and DC conductivity measurements, were performed to characterize these polymer electrolytes. Based on these electrolytes, electrochemical cells were fabricated with anode/polymer electrolyte/cathode configurations. The discharge characteristics of the cell were studied for a load of 100 kΩ.

## 2. Experimental

### 2.1 Preparation of plasticized nanocomposite polymer electrolytes

**Figure 1** illustrates the preparation method of solid polymer electrolytes. **Table 1** shows various compositions of filler incorporated GPEs. Polyacrylonitrile (PAN) with a molecular weight of 1,50,000 g/mol was used as the host polymer and sodium fluoride (NaF) was used as the dopant salt. Ethylene carbonate (EC) was used as a plasticizer in the electrolyte and aluminum oxide ($Al_2O_3$) was used as a nanofiller. Solid polymer electrolytes were prepared by mixing PAN and NaF salt in dimethyl formamide (DMF). The solution thus obtained was.

stirred in a magnetic stirrer for approximately 4 hours until we obtained a homogeneous translucent gel after the ethylene carbonate (EC) plasticizer was added to this solution and stirred for approximately 2 hours. After that, the Al2O3 nanofiller was added to this solution and stirred for 48 hours to disperse the nanofiller homogenously in the polymer matrix [2]. The resulting homogeneous mixture was then cast onto polypropylene dishes and the solvent was allowed to evaporate at room

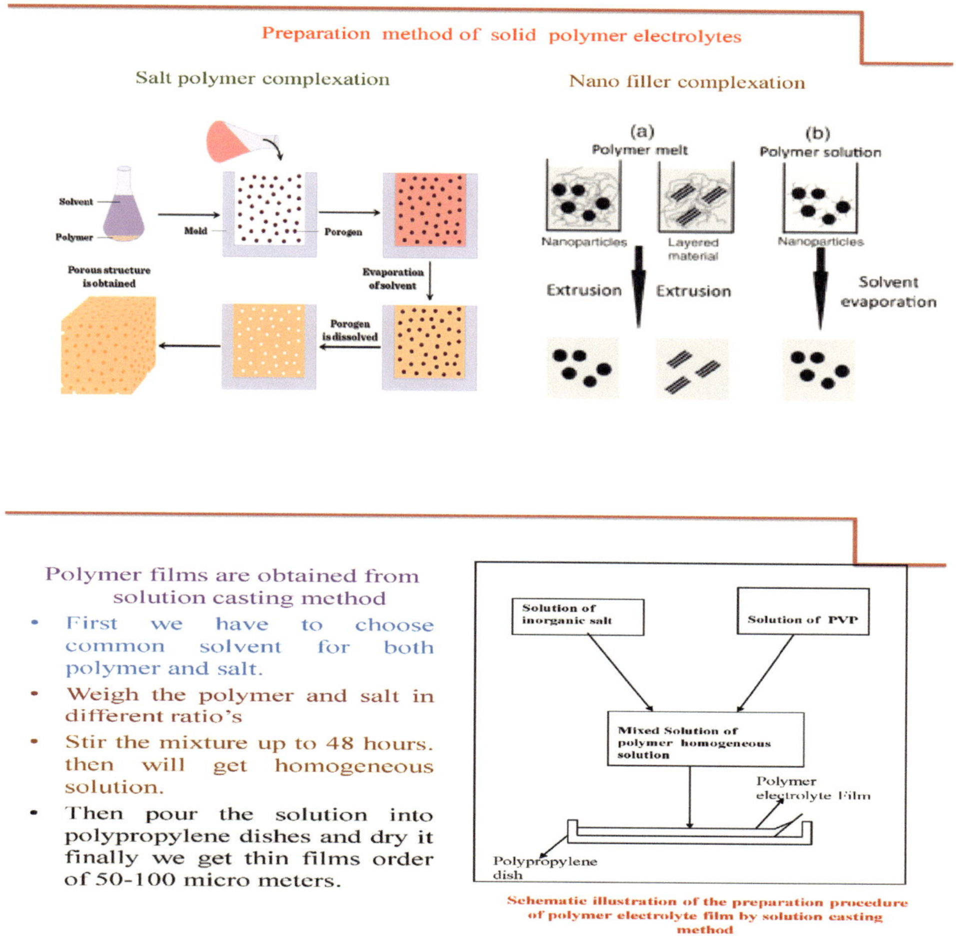


**Figure 1.**
*Preparation method of solid polymer electrolyte films.*

| Sample code | Composition (mol %) | | | |
|---|---|---|---|---|
| | PAN | NaF | DMF + EC | $Al_2O_3$ |
| PURE | 700 mg | 300 mg | 20 ml + 880 mg | 0 |
| 70PAN:30NaF:1 wt% | 700 mg | 300 mg | 20 ml + 880 mg | 1 |
| 70PAN:30NaF:2 wt% | 700 mg | 300 mg | 20 ml + 880 mg | 2 |
| 70PAN:30NaF:3 wt% | 700 mg | 300 mg | 20 ml + 880 mg | 3 |
| 70PAN:30NaF:4 wt% | 700 mg | 300 mg | 20 ml + 880 mg | 4 |

**Table 1.**
*Various compositions of filler incorporated gel polymer electrolytes [5].*

temperature. After the solvent had completely evaporated, the films were peeled off from the polypropylene dishes and pressed under a membrane hot press at a temperature of 40°C. The pressure applied by this method was 3.5 torr/cm$^2$, and we obtained flexible and self-standing polymer electrolytes. The samples consisted of 1 wt% $Al_2O_3$ (A1), 2 wt% $Al_2O_3$ (A2), 3 wt % $Al_2O_3$ (A3), and 4 wt% $Al_2O_3$ (A4). The prepared electrolyte membrane systems had a thickness of approximately 128 μm.

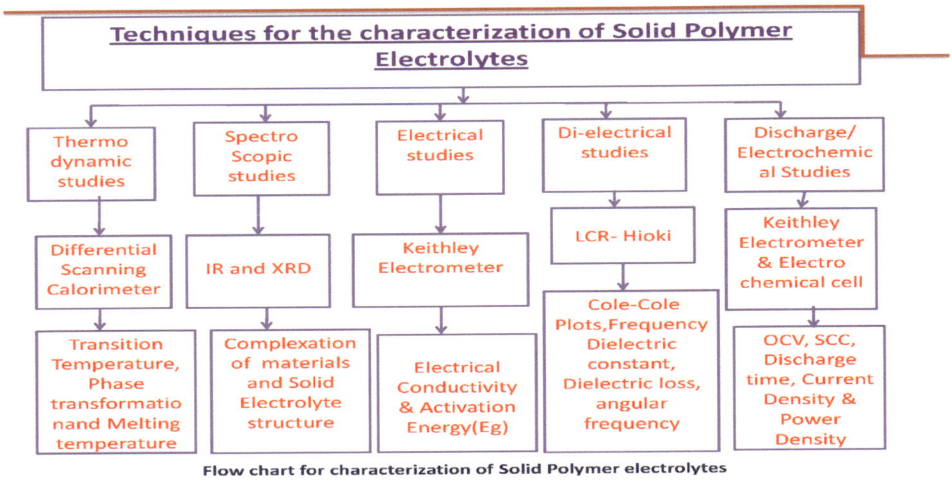


**Figure 2.**
*Flow chart techniques for the characterization of solid polymer electrolytes.*

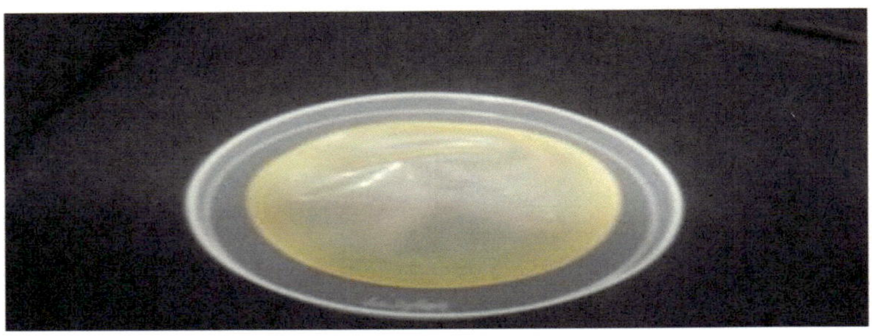

**Figure 3.**
*Gel polymer electrolyte film based on nanofillers ($Al_2O_3$).*

## 2.2 Materials characterization

**Figure 2** shows a flowchart for the characterization of solid polymer electrolytes. In the present work, the prepared films were characterized by energy dispersive x-ray spectrometry (EDS) to determine the chemical composition in wt%. The UV–Vis spectrum was recorded by a Hewlett-Packard HP8452A diode array spectrometer. The mechanical properties of the prepared polymer electrolytes were determined using a Universal Tensile Machine (Instron Model 5565, Canada) with a constant crosshead speed of 10 mm/min. The sample dimensions were 25 mm × 40 mm × 0.1 mm. Additionally, DSC thermograms were recorded to measure the glass transition temperature and melting temperature of PAN-based electrolytes by using a Mettler instrument. The samples were heated sequentially from 50–360°C. Finally, transport characteristics and discharge characteristics, such as transference number, open circuit voltage (OCV), short circuit current (SCC), and power density, were determined. at a constant load of 100KΩ. **Figure 3** illustrates a GPE film based on a $Al_2O_3$ nanofiller.

## 3. Results and discussion

**Figure 4** reveals the energy dispersive spectra of the prepared samples. From these spectra, the chemical composition of composite materials can be explained.

**Energy Dispersive Spectroscopy (EDS):**

| ELEMENT | Weight% | Atomic% |
|---|---|---|
| C | 66.89 | 71.18 |
| N | 23.62 | 21.55 |
| O | 7.55 | 6.03 |
| F | 1.46 | 0.98 |
| Na | 0.42 | 0.24 |
| Al | 0.06 | 0.03 |
| Total | 100.00 | 100.00 |

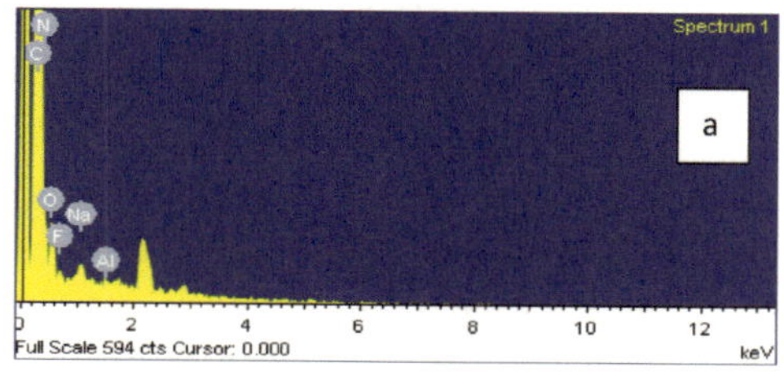


| ELEMENT | Weight% | Atomic% |
|---|---|---|
| C | 56.71 | 61.76 |
| N | 30.73 | 28.69 |
| O | 8.37 | 6.84 |
| F | 2.93 | 2.02 |
| Na | 0.94 | 0.53 |
| Al | 0.32 | 0.16 |
| Total | 100.00 | 100.00 |

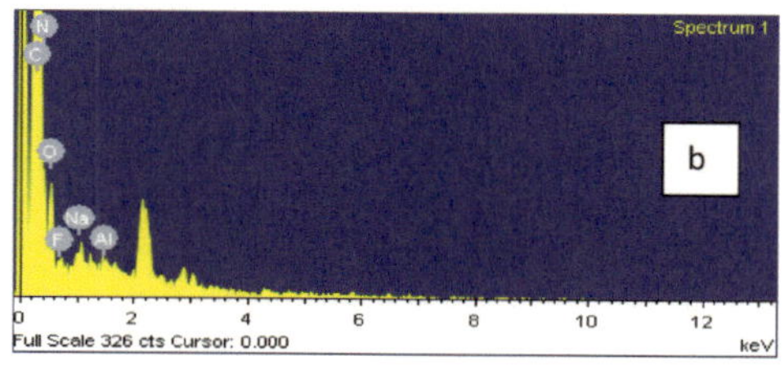


| ELEMENT | Weight% | Atomic% |
|---|---|---|
| C | 63.37 | 67.91 |
| N | 26.56 | 24.41 |
| O | 7.47 | 6.01 |
| F | 1.96 | 1.33 |
| Na | 0.45 | 0.25 |
| Al | 0.19 | 0.09 |
| Total | 100.00 | 100.00 |

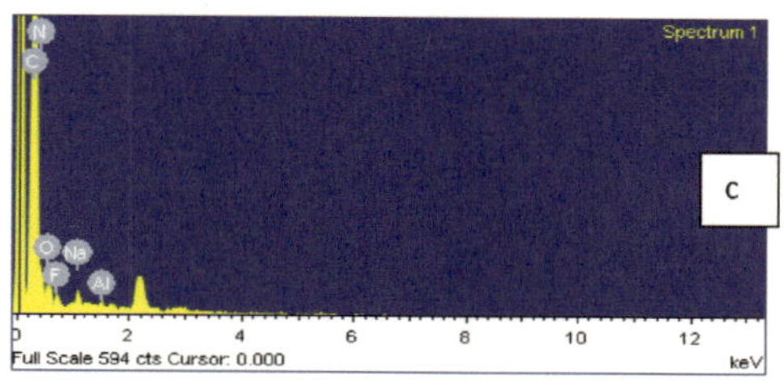


| ELEMENT | Weight% | Atomic% |
|---|---|---|
| C | 65.32 | 69.54 |
| N | 26.26 | 23.97 |
| O | 7.03 | 5.62 |
| F | 0.95 | 0.64 |
| Na | 0.31 | 0.17 |
| Al | 0.14 | 0.07 |
| Total | 100.00 | 100.00 |

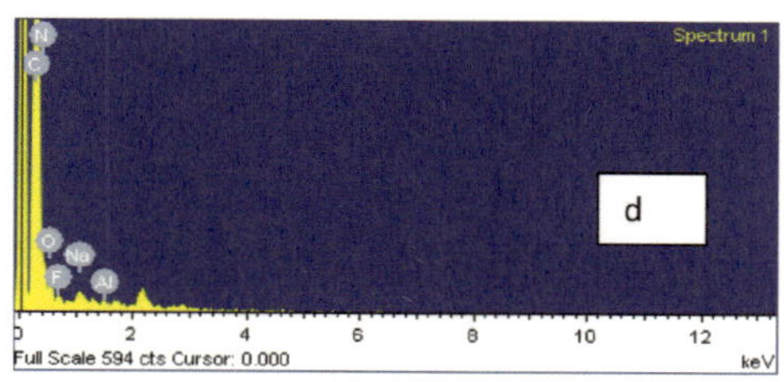


**Figure 4.**
*[a-d]. EDS spectra of PAN:NaF (70:30) + (1–4) Wt% $Al_2O_3$ nanocomposite gel polymer electrolyte films.*

The analysis indicates the presence of carbon (C), nitrogen (N), sodium (Na), fluoride (F), oxygen (O), and aluminum (Al) with chemical elements in the GPE of PAN + NaF (70:30) + Wt% $Al_2O_3$. EDS can be used to estimate the relative abundance and accuracy of quantitative analysis. This analysis explains the structural properties of prepared samples, and these properties may affect various factors [6–8]. With increasing nanofillers in the polymer composite, the atomic nature of the samples can be modified, and the results are represented in the above tabular forms.

## 4. UV: Visible spectroscopy

**Figure 5** shows the UV–Vis spectra of PAN with different wt. % ratios of NaF salt and $Al_2O_3$ nanoparticles at room temperature. The UV–Vis spectrum was recorded by a Hewlett-Packard HP8452A diode array spectrometer. The structural effect of salt and nanoparticles on the conductivity was also confirmed by UV–Vis spectroscopy. Optical absorption, particularly studying the shape and shift of the absorption edge, is a useful technique for understanding the basic mechanism of optically induced transitions in crystalline and noncrystalline materials. UV–Vis

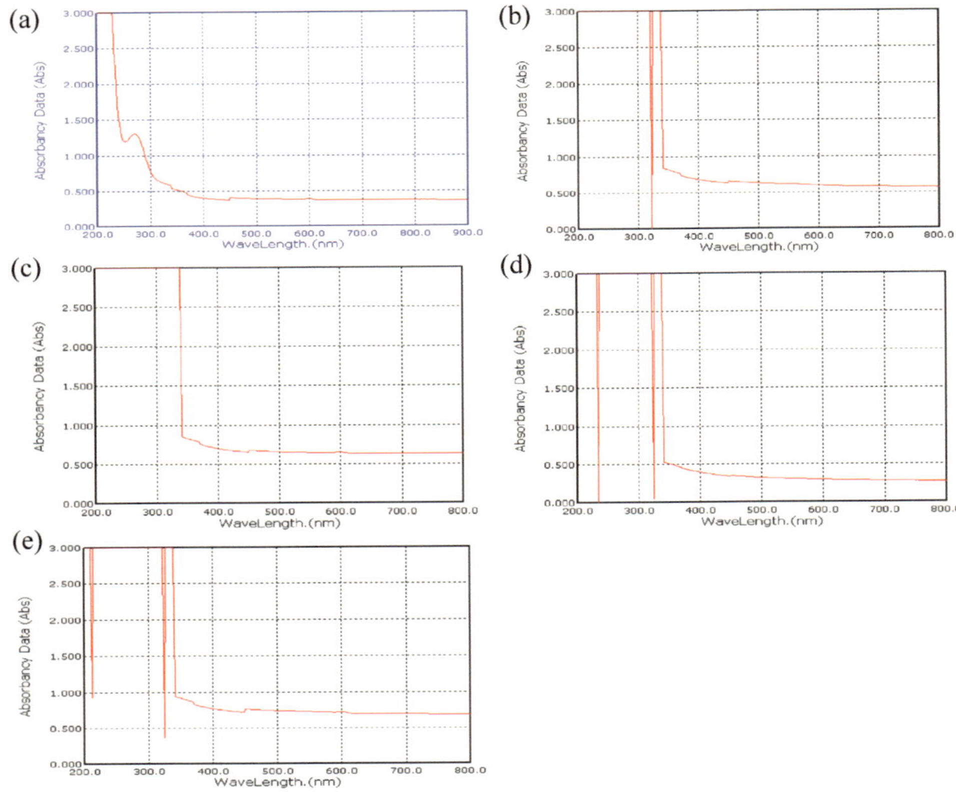


**Figure 5.**
*(a-e): UV–VIS spectroscopic images of PAN:NaF (70:30) and different ratios of nanoparticle gel polymer electrolyte films at room temperature.*

spectroscopy is used to identify inorganic complexation of molecules and their qualitative and quantitative measurements [9]. It is also used to identify the energy band gap values of the materials in the transmitting radiation. At an energy level, a photon is absorbed in its orbit. When an electron jumps from a lower energy level to a higher energy level. Transitions take place in a band gap energy as it rises in the absorption process called the absorption edge, where the optical band gap energies are determined [10–12]. The absorption rate is slightly changed by increasing the salt ratio of solvents and nanoparticles. The optical band gap of the polymer electrolytes was determined using UV–Vis spectra. It can be determined by

$$E = hc / \lambda \quad (1)$$

where h is Planck's constant ($6.626 \times 10^{-34}$ joules sec), c is the light velocity ($3 \times 10^{8}$ meters/sec), and λ is the cutoff wavelength. **Table 2** gives the wavelength values from the UV–Visible spectra. It is cleared that the optical energy band.

## 4.1 DSC characteristics

The Mettler instrument was calibrated with indium and zinc standards and the analyses were conducted under a nitrogen flow rate of ca. 20 mL/min. The sample was heated sequentially from 50–360°C. The change in transition temperature caused by the incorporation of nanofiller $Al_2O_3$ and plasticizer into the PAN+NaF complex was studied by DSC analysis. **Figure 6** shows the DSC thermograms of 70PAN:30NaF and 1–4 wt% $Al_2O_3$.

| Polymer electrolyte | Planck's constant (h) | Light velocity (C) | Wavelength (Å) | Optical energy band gap in (ev) |
|---|---|---|---|---|
| Pure PAN | $6.626 \times 10^{-34}$ joules sec | $3 \times 10^{8}$ meter/sec | 350.12 | 3.54842 |
| 70PAN:30NaF:1 wt% | $6.626 \times 10^{-34}$ joules sec | $3 \times 10^{8}$ meter/sec | 340.0 | 3.65404 |
| 70PAN:30NaF:2 wt% | $6.626 \times 10^{-34}$ joules sec | $3 \times 10^{8}$ meter/sec | 352.0 | 3.52787 |
| 70PAN:30NaF:3 wt% | $6.626 \times 10^{-34}$ joules sec | $3 \times 10^{8}$ meter/sec | 360.0 | 3.44803 |
| 70PAN:30NaF:4 wt% | $6.626 \times 10^{-34}$ joules sec | $3 \times 10^{8}$ meter/sec | 350.0 | 3.54947 |

**Table 2.**
*Optical energy band gap for PAN-based polymer electrolytes.*

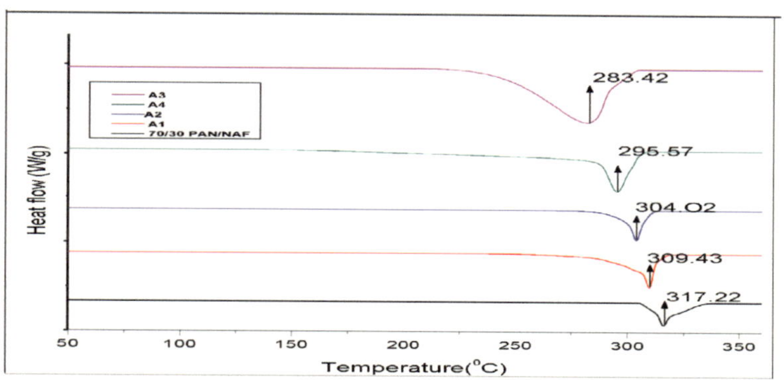


**Figure 6.**
*DSC thermographs of the 70PAN:30NaF, A1, A2, A3, and A4 samples.*

**Table 3** summarizes melting temperature (Tm) and the corresponding heat enthalpy (ΔHm) and the percentage of crystallinity ($\chi_c$) of the prepared polymer electrolytes. The percentage of crystallinity was calculated by

$$\chi_c = \Delta H_m * / 398.6 jg^{-1} \tag{2}$$

where ΔHm* is the heat enthalpy of the polymer electrolytes, PAN has $T_m$,$T_g$ and $\Delta H_m$ of 317°C, 107°C, and 398.6 $Jg^{-1}$respectively. From the table, it can be concluded that the

| Sample | $T_m$(°C) | $\Delta H_m^*$((j/g) | Crystallinity (%) |
|---|---|---|---|
| Pure (Filler free) | 317.22 | 398.6 | 98 |
| 70PAN:30NaF:1 wt% | 309.43 | 298.3 | 87 |
| 70PAN:30NaF:2 wt% | 304.02 | 243.6 | 74 |
| 70PAN:30NaF:3 wt% | 283.42 | 132.5 | 60 |
| 70PAN:30NaF:4 wt% | 295.57 | 175.4 | 63 |

**Table 3.**
*$T_m$, $\Delta H_m^*$, $\chi_c$ of 70PAN:30NaF, A1, A2, A3, A4 samples.*

incorporation of NaF salt and $Al_2O_3$ nanopowder into the polymer blend matrix decreases the $T_m$ value, and the minimum Tm value is 283.42°C for the 3 wt.% $Al_2O_3$ nanopowder content. This observation suggests an increase in the crystallinity of the complexes because of the presence of excess nanopowder. It is also evident from conductivity studies that the conductivity of the complexes increases with increasing salt concentration and decreases for greater concentrations of nanopowder due to the formation of ion clusters. When sodium fluoride salt was added to the host polymer PAN, the results obtained indicated that the filler–polymer interaction influenced the speed of sodium ions in the polymer chain. This is in good agreement with TGA results [5, 13–16].

## 4.2 Tensile test: 70PAN-30NaF-(1–4 wt%) $Al_2O_3$ composite

To effect the practical use of a polymer electrolyte, the electrolyte must remain structurally stable during manufacturing, cell assembly, storage, and usage, prevent flow from occurring within the cell to prevent self-discharge, and be easy to prepare in a repeatable manner. That is, mechanical strength is also an important factor when manufacturing polymer electrolytes. As noted earlier, incorporating additives such as ceramic powder can strengthen the dimensional stability of electrolyte membranes [17–20].

The addition of nanoparticles increased both the modulus and the strength of the polymer nanocomposites. Additionally, the toughness (area under the stress–strain curve before rupture) increased significantly. **Figure 7** shows the tensile strength (the maximum stress in the stress–strain curve, MPa) and Young's modulus (the slope of the stress–strain curve in the low strain region) as a function of nanoparticle volume content. Both the tensile strength and Young's modulus increased with increasing functionalized particle loading. Compared to the pure polymer, the strength and Young's modulus of the 4 wt% filled nanocomposite sample increased by approximately 99%. The imagination of the nanoparticles was observed to have a greater effect on the Young's modulus. Moreover, the relatively uniform distribution of $Al_2O_3$ particles and decrease in interparticle distance with increasing particle loading in the matrix results in polymer nanocomposites having increased resistance to indentation. For a given volume fraction, nanoparticles are much closer to each other than microparticles in the matrix, and hence nanoparticles will more strongly resist the penetration of the indentation in the matrix [21]. This results in higher microhardness for nanocomposites than that of polymers at a constant volume fraction of particles.

## 4.3 DC conductivity studies

### *4.3.1 Conductance spectra for PAN:NaF gel polymer electrolytes*

**Figure 8** shows the ionic conductivity in addition to NaF salt with the host polymer PAN. The conductivity increased with the enhancement of NaF salt from 10 to

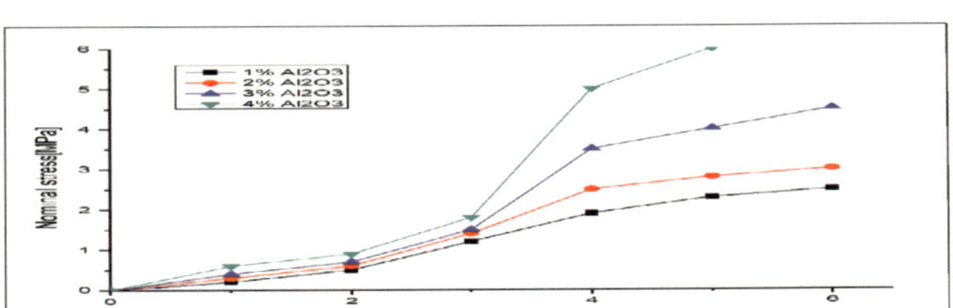


**Figure 7.**
*Tensile test of 70PAN-30NaF-(1–4 wt%)$Al_2O_3$Composite.*

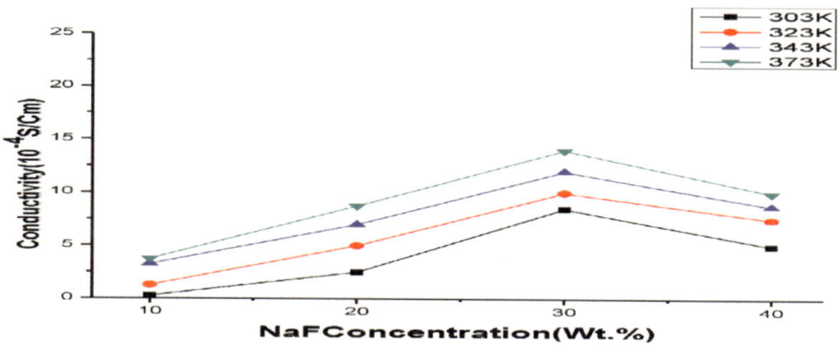


**Figure 8.**
*Composition dependence of conductivity in the PAN+NaF system.*

30 wt%. For 40 wt% NaF salt, the ionic conductivity was decreased. **Table 4** gives the ionic conductivity values. With the doping of NaF salt, the number of free ions also increased in the host polymer, resulting in enhanced conductivity. At higher concentrations, the enhanced viscosity of the polymer reduced the ionic conductivity. For 30 wt%, NaF shows a higher conductivity of 1.82 x $10^{-4}$ S $cm^{-1}$ due to its less crystalline nature compared to other compositions [16–18].

Conductivity studies were carried out for all the prepared polymer electrolytes to understand the conduction mechanism. **Figure 9** shows the ionic conductivity versus nanofiller ($Al_2O_3$) concentration of the polymer electrolyte PAN+NaF complexed system with varying weight percentages of nanofiller in the temperature range 303 K to 373 K. **Table 4** presents the conductivity data at room temperature and at 373 K. We conclude that the conductivity increases as the nanofiller content increases up to 4 wt% due to the high.

Amorphous nature of the polymer electrolyte, which provides more free mobile protons (carriers), thus giving rise to higher conductivity [22].

The enhancement in conductivity is not only due to the increment of mobile charge carriers but also due to ethylene carbonate (EC), which allows greater dissolution of the electrolyte salt and nanofiller, resulting in an increased number of charge carriers and hence an increase in conductivity. The maximum value of conductivity obtained at room temperature is 4.82 × 10–3 S $cm^{-1}$. This conductivity value is 10 orders greater than that of pure PAN ($10^{-14}$ S $cm^{-1}$), as reported by Pan and Zou [23]. The conductivity increases with the incorporation of nanofiller wt %, which might be due to ion pair or aggregate formation [24]. The polymer electrolytes lead to an increase in the viscosity of the polymer electrolyte film due to the incorporation of a large amount of salt and nanofiller content, which reduces

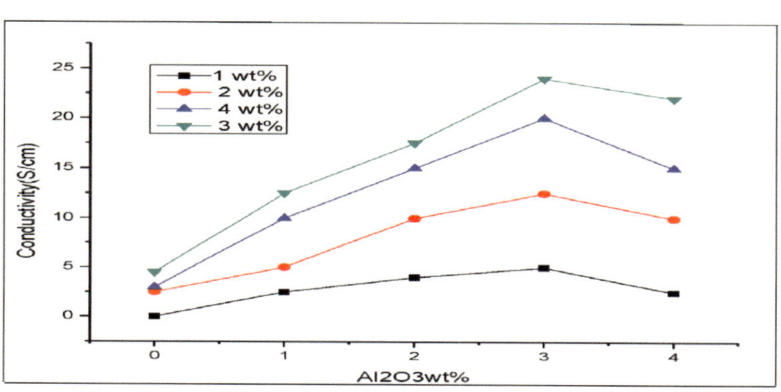


**Figure 9.**
*Ion conductivity (σ) of polymer electrolyte films as a function of wt% of $Al_2O_3$ concentration.*

| GPEs | Ionic Conductivity at 303 K | 373 K | Activation energy(eV) | Transference Number | |
|---|---|---|---|---|---|
| | | | | tion | tele |
| PAN: NaF (90:10) | 5.46 x $10^{-6}$ | 2.85 x $10^{-5}$ | 0.47 | 0.965 | 0.04 |
| PAN:NaF (80:20) | 1.62 x $10^{-5}$ | 2.42 x $10^{-5}$ | 0.32 | 0.972 | 0.014 |
| PAN:NaF (70:30) | 1.82 x $10^{-4}$ | 2.96 x $10^{-3}$ | 0.25 | 0.989 | 0.03 |
| PAN:NaF (60:40) | 2.64 x $10^{-5}$ | 1.32 x $10^{-5}$ | 0.28 | 0.969 | 0.02 |

**Table 4.**
*DC conductivity, $E_a$, ionic transference numbers of various compositions of PAN:NaF gel polymer electrolytes.*

proton transportation and impedes the mobility of charge carriers, resulting in a decrease in proton conductivity.

**Figure 10** Illustrates the variation of ionic conductivity with temperature for different wt% of $Al_2O_3$. The ionic conductivity was calculated using the formula given in Eq. 3.

$$\sigma = t / AR_b \quad (3)$$

where t and A represent the thickness and the area of the electrolyte specimen, respectively [25]. $R_b$ is the bulk resistance of the electrolyte obtained from the complex impedance measurement. The ionic conductivity depends on the overall mobility of ion species present in the electrolyte and the polymer, which is determined by the free volume made by filler and plasticizer around the polymer chain. In the present study, conductivity enhancement was observed when the $Al_2O_3$ filler and EC plasticizer were incorporated into the polymer salt system. The polymer electrolytes considered in the present study exhibited an Arrhenius type of conduction [26]. The polymer electrolyte containing 3 wt% nanofiller showed an ionic conductivity of 5.96 × $10^{-3}$ $Scm^{-1}$.

### *4.3.2 Transference number measurements*

To measure the conductivity of prepared polymer electrolytes, the transfer numbers play a vital role. To determine transfer numbers, Wagner's polarization

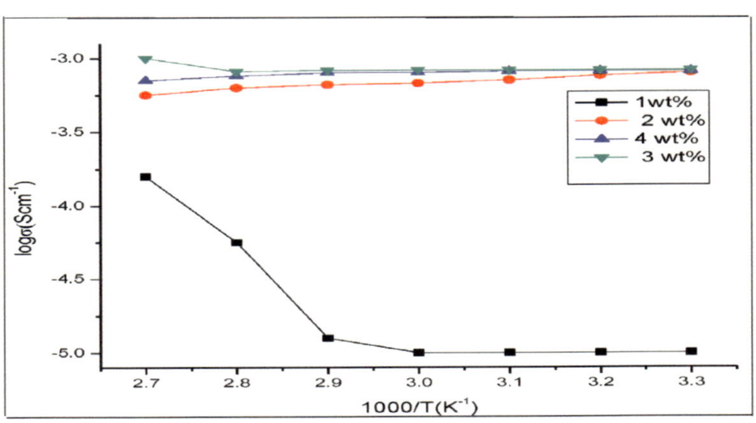


**Figure 10.**
*Log σ versus 1000/T of PAN+NaF complexed films for different wt% of $Al_2O_3$.*

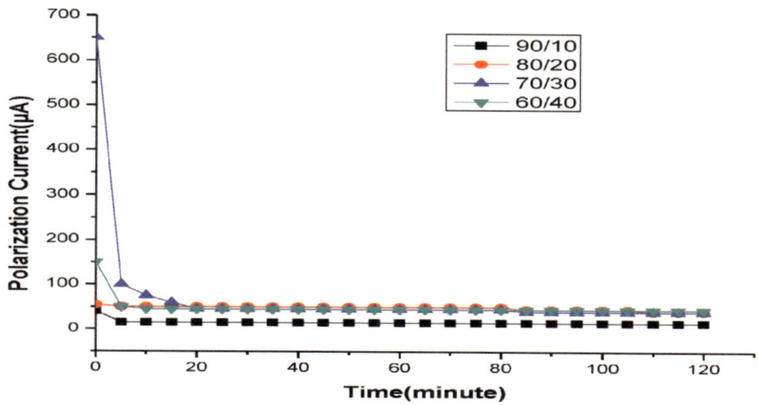


**Figure 11.**
*Polarization current vs. time plot for the PAN:NaF polymer composite at 303 K.*

method can be used. The prepared PAN:NaF sample was placed between two silver electrodes. After the circuit was closed, DC voltage was applied to the sample to measure the polarization current. **Figure 11** shows the final stabilized current values. The polarization current was decreased with respect to increasing time due to predominantly ions [19]. **Table 4** shows the transfer numbers for 10 to 40 wt% NaF doped with PAN as 0.965, 0.972, 0.989, and 0.969.

## 5. Working principle and design of ideal solid polymer batteries

An ideal polymer battery exhibiting a polymer electrolyte membrane, a positive electrode and a negative electrode. The positive electrode is flexible with elastic materials and the negative electrode consists of metal foil. During the recharging process, the direction of flow of electrons takes place in the opposite (anode-to-cathode) direction. Designing new solid-state batteries that have better performance over the life cycle also depends on the outstanding operation of electrode–electrolyte interfaces, as shown in **Figure 12**.

Most polymeric materials are insulators that do not conduct electricity in the medium through ions or electrons. Some of the polymers are "immobile solvents" at which the salt and the polymer are completely mixed and act as a polymer electrolyte [27–31]. Compared to solid electrolytes, polymer electrolytes have both cations and anions in the mobile state. Due to this, polymer batteries designed to be rechargeable face several challenges. Fabrication of polymer batteries is an important method to pass ions from the anode to the cathode. Here, sodium metal was considered an anode and carbon mixed with iodine was taken as a cathode material.

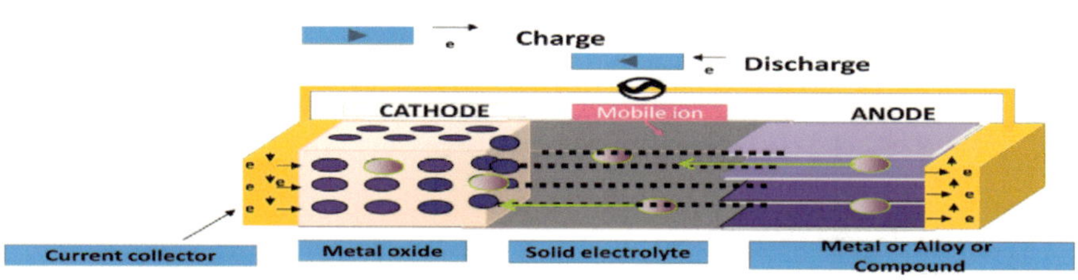


**Figure 12.**
*Schematic diagram of a solid polymer battery.*

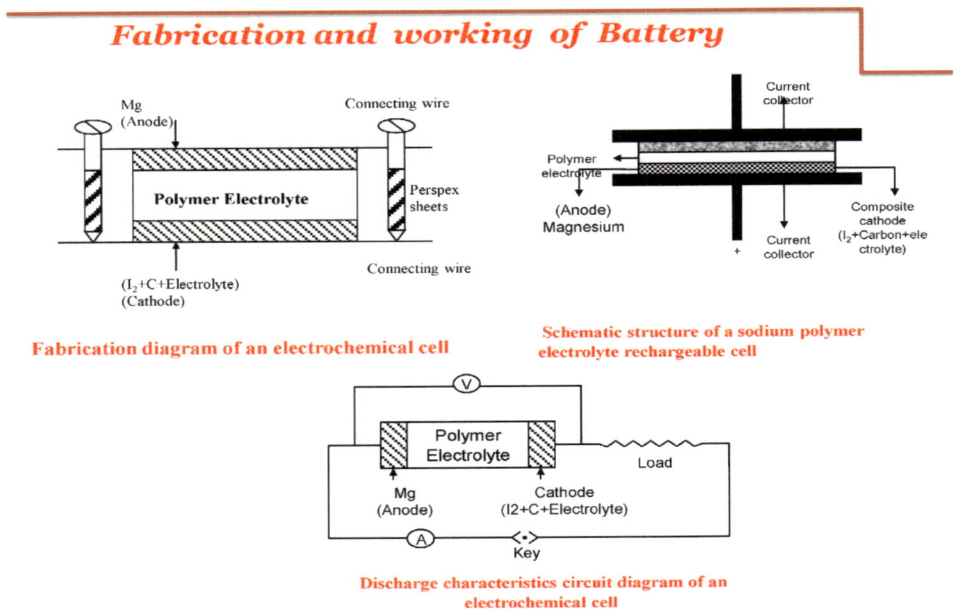


**Figure 13.**
*Fabrication and operation of a solid-state battery.*

For some PEO N:$LiCF_3SO_3$, mobilization takes place at the anode due to anions, and it reacts with Li + cations, which form a thin layer of lithium salt and act as a conducting medium.

$$Li^+ + X^- LiX$$

The results show that the battery resistance is increased, and Li + ions decrease the capacity. Polarization takes place in an electrolyte due to the mobility of ions in the battery system with respect to negative ions.

### 5.1 Discharge characteristics of a solid-state battery

Solid-state batteries provide well-contained energy conversion devices, which greatly contribute to the needs of humankind. Zero-emission vehicles of the future will be battery powered only. Many nonpolluting energy conversion devices, such as photovoltaic systems, require the concomitant use of rechargeable batteries for energy storage. Batteries may be considered storehouses for electrical energy [1–4, 6–10]. The size of a battery ranges from a tiny coin to that of a large house. Tiny coin and button-sized cells are used for electronic applications requiring only small capacity. Liter-container sized batteries are commonly used in motor vehicles for starting, lighting, and ignition purposes. The basis for battery technology is that the chemical energy derived from the chemical reactions in the battery is transformed into electrical energy. **Figure 13** shows the fabrication and operation of a solid-state battery.

## 6. [70PAN + 30NaF] gel polymer electrolyte system

### 6.1 Discharge characteristics

A proton battery using the membrane with the maximum ionic conductivity was constructed with sodium (Na) as an anode material and a mixture of iodine ($I_2$), carbon, and a piece of GPE taken as the cathode electrode between the 70PAN:30NaF electrolyte film [5]. **Figure 10** represents the discharge characteristics of the electrochemical cell with an applied load of 100 k Ω at room temperature. From these results, the OCV of the cell was observed for 118 hours and plotted against time, as shown in **Figure 14**. A stabilized voltage of 2.78 V was obtained from the cell. The discharging curve initially decreases with the voltage of the cell, which may be due to the polarization effect. **Table 5** lists the OCV and discharge time for the cell [32–37].

A solid-state electrochemical cell was fabricated with an anode Na/PAN:NaF (70:30)/cathode ($I_2$ + C + electrolyte+$Al_2O_3$). The thickness of both the electrodes is 1 mm. The surface area and thickness of the PAN+NaF + $Al_2O_3$ GPE were 1.28 cm2 and 128 μm, respectively. The discharge characteristics of the cell for a constant load of 100 kΩ were evaluated at room temperature as shown in **Figure 14**.

The initial sharp decrease in the voltage in these cells may be due to polarization and the formation of a layer of sodium salt at the electrode–electrolyte interface [38–41]. Cell parameters such as OCV, SCC, current density, power density, energy density, and discharge capacity have been evaluated in the highest conducting GPE

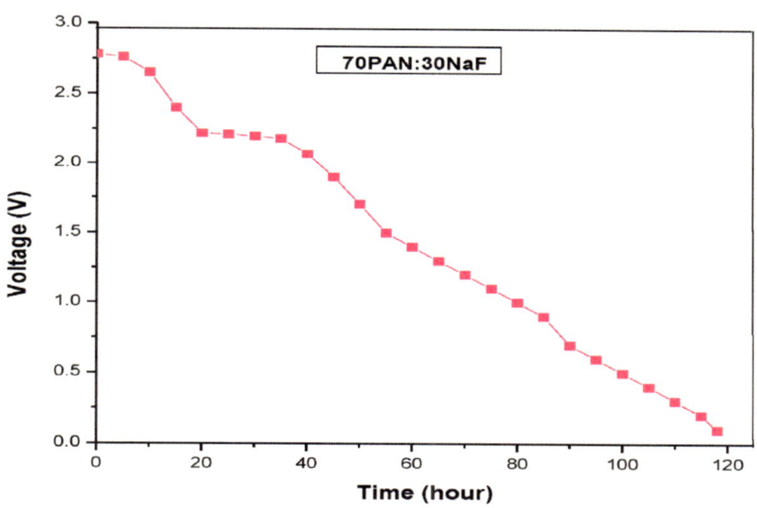


**Figure 14.**
*Discharged curve of the solid-state cell configuration of Na/PAN:NaF (70:30)/cathode ($I_2$+ C + electrolyte+ $Al_2O_3$) at a load of 100 kΩ.*

| Cell parameters | 70PAN:30NaF |
|---|---|
| Area of the electrolyte ($cm^2$) | 1.30 |
| Cell weight (gm) | 1.5 |
| OCV (V) | 2.78 |
| SCC (μA) | 1.2 |
| Discharge time for plateau region (h) | 118 |
| Density of power (mW/kgg) | 2.94 |
| Density of energy (mWh/kg) | 265.4 |
| Density of current (mW/kg) | 2.32 |
| Capacity of discharge (μA $h^{-1}$) | 152.4 |

**Table 5.**
*Cell parameters of PAN:NaF (70:30) gel polymer electrolytes.*

| Cell parameters | 70PAN:30NaF:3 wt% Al2O3 |
|---|---|
| Area of the electrolyte ($cm^2$) | 1.20 |
| Cell weight (g) | 1.64 |
| OCV (V) | 3.20 |
| SCC (mA) | 1.6 |
| Discharged time for plateau region (h) | 142 |
| Density of power (W/kgg) | 2.94 |
| Density of energy (Wh/kg) | 324.7 |
| Density of current (Mw/kg) | 2.35 |
| Capacity of discharge (μA $h^{-1}$) | 168.2 |

**Table 6.**
*Cell parameters of the PAN:NaF + $Al_2O_3$/($I_2$+ C + electrolyte) polymer electrolyte battery.*

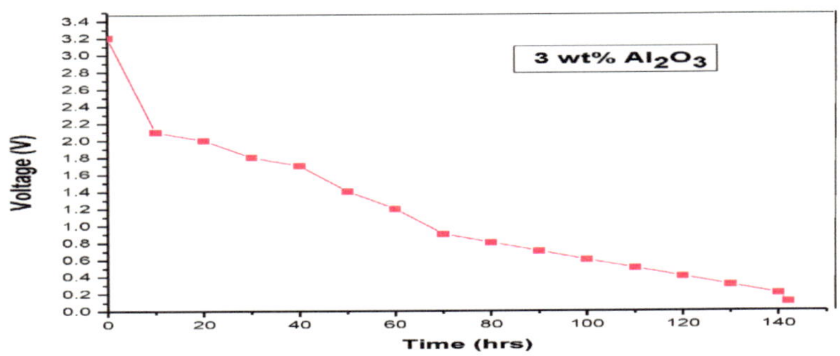


**Figure 15.**
*Discharge characteristics of PAN:NaF complexed with 3 wt% $Al_2O_3$ nanofiller polymer electrolyte electrochemical cell (load =100 kΩ).*

system PAN:NaF (70:30) + nanofiller ($Al_2O_3$) in this electrochemical cell. **Table 6** shows the obtained data. The current density is calculated using the SCC value and area of the cell. The power density value is obtained by taking the OCV and weight of the cell into consideration. The energy density value is calculated by evaluating the time taken for the plateau region [5]. From **Table 6**, it is obvious that the cell with the composition PAN:NaF (70:30) + nanofiller $Al_2O_3$ exhibits better performance [26]. It is confirmed that gel-state cell parameters are better than the earlier reported sodium-based polymer electrolyte cell system (**Figure 15**) [27–31].

## 7. Conclusions

Proton-conducting GPEs consisting of NaF salt dissolved in a plasticizing solvent of EC and DMF, immobilized in a host polymer 70PAN + 30NaF + (1–4 wt%) $Al_2O_3$ (nanofiller) were synthesized and characterized. The complexation of the salt and nanofiller with the polymer was confirmed by ED. UV–Vis light absorption reveals that the chemical structure of the polymer is identical to that of the polymer formed electrochemically. The various absorption rates given for different wavelengths and optical energy band gaps were determined. The tensile test of the prepared samples was verified by a stress–strain graph. The tensile strength was increased to 5.9 MPa for the polymer electrolyte with 3 wt% nanocomposite compared to the filler-free electrolyte. A decrease in the degree of crystallinity and an increase in the amorphous nature were observed, while an increase in conductivity was observed with increasing nanofiller concentration and temperature. The heat flow observed by DSC technique that the transference data indicates the conduction in the polymer electrolyte is predominantly due to ions rather than electrons. Using a PAN:NaF (70:30) GPE system, a solid–state battery (Na/PAN:NaF (70:30) + EC + DMF/($I_2$ + C + electrolyte)) was fabricated, and its discharge characteristics were studied. These results were found to be comparable with existing results.

## Acknowledgements

The author thanks Vice Chairman of Dayananda Sagar Institutions, Bangalore Dr. D. Premachandra Sagar; Vice-Chancellor Prof. KNB Murthy; and Dean Dr. A. Srinivas SOE, Dayananda Sagar University for constant support and encouragement. I especially thank my beloved student Narasimha Rao Maragani, who supported and helped me with the research work.

## Author details

Vijaya Kumar Kambila
Department of Physics, Dayananda Sagar University, Bangalore, India

*Address all correspondence to: drkambilavk@gmail.com

## References


[1] Jaipal Reddy M, Siva Kumar J, SubbaRao UV, Chu PP. Structural and ionic conductivity of PEO blend PEG solid polymer electrolyte. Solid State Ionics. 2006;**177**:253-256

[2] Maragani Narasimhara O, Vijayakumar K. Structural and ionic conductivity studies on plasticized PAN-sodium fluoride polymer electrolytes for electro chemical cell applications. Iranian Journal of Materials Science and Engineering. 2017;**14**(4):1-10

[3] Sreepathi Rao S, Rao KVS, Shareefuddin M, Subbarao UV, Chandra S. Ionic conductivity and battery characteristics studies on PEO+$AgNO_3$ polymer electrolyte. Solid State Ionics. 1994;**67**:331-334

[4] Narasimharao M, Vijayakumar K. "AC conductivity and thermal characterization of PAN doped gel polymer electrolytes for polymer battery applications." vol. 10 | No. 2 |665 -672 | April - June | 2017 ISSN: 09741496 | e-ISSN: 0976-0083 | CODEN: RJCABP. Rasayan Journal of Chemistry. 2017;**10**(2):665-672

[5] Krishnajyothi N, Vijaykumar K. Ionic conductivity and battery characteristics studies of a new PAN – based Na+ ion conducting gel polymer electrolyte system. IJP-2015

[6] Sarojini S, Anjalai C. Synthesis and structural characterization studies on solid polymer electrolyte system with magnesium triflate as host salt, EC as plasticizer and MgO as nanofiller. Chemical Science Transactions. 2016;**5**(3). ISSN:2278-3458

[7] Narasimharao M, Vijayakumar K. "Ion conducting gel polymer electrolyte based on poly (acrylonitrle) complexed with (NaF+ $ZrO_2$) application as an electrochemical cell." vol. 10 | No. 4 |1128 -1136 | October - December | 2017 ISSN: 0974-1496 | e-ISSN: 0976-0083 | CODEN: RJCABP. Rasayan Journal of Chemistry. 2017;**10**(4):1128-1136

[8] Rajendran S, Mahalingam T, Kannan R. Experimental investigations on PAN-PEO hybrid polymer electrolytes. Solid State Ionics. 2000;**130**:143-148

[9] Narasimharao M, Vijayakumar K. AC conductivity and thermal studies of PAN-NaF doped gel polymer electrolytes for solid-state battery applications. International Journal of Chemical Concepts. 2017;**3**(3):27284. ISSN: 2395-4256

[10] Vijaya KK, Suneetha Sundari G. Studied of (PEO+$KHCO_3$) solid electrolyte system and its application as an electrochemical cell. Journal of Engineering Science and Technology. 2010;**5**:130-139

[11] Ramesh Babu J, Vijaya Kumar K. Studies on structural and electrical properties of $NaHCO_3$ doped PVA films for electrochemical cell applications. ChemTech;**7**(1):171-180

[12] Narasimharao M, Vijayakumar K. Structural, thermal and battery characteristic properties of $NH_4CF_3SO_4$ doped pan films for electrochemical cell applications. International Journal of ChemTech Research. 2016;**9**(05):432-438. ISSN: (Online): 2455-9555

[13] Wagner JB, Wagner C. Electrical conductivity measurements on cuprous halides. Journal of Chemical Physics. 1957;**26**:1597-1601

[14] Narasimharao M, Vijayakumar K. "Structural and a.C. conductivity studies of (PAN+ NaF) gel polymer electrolyte system with $ZrO_2$ nanofiller for an electro chemical cell application." vol. 10 | No. 4 |1218 -1225 | October - December | 2017 ISSN: 0974-1496 | e-ISSN: 0976-0083 | CODEN: RJCABP. Rasayan Journal of Chemistry. 2017;**10**(4):1218-1225

[15] Perkumpus H-H. UV–Visible Spectroscopy and its Applications. New York: Springer–verlag; 1992

[16] Tomar R, Sharma CR. Studies on effect of doping alumina nanoparticles in ION conducting polymer nanocomposites. Tomar. 2013;**2**(9). ISSN: 2277-9655

[17] Yadav VS, Sahu DK, Singh Y, Dhubkarya DC. The Effect of Frequency and Temperature on Dielectric Properties of Pure Poly Vinylidene Fluoride (PVDF) Thin Films. Vol. III. IMECS; 2010. ISBN: 978-988-18210-5-8

[18] Krishnajyothi N, Vijaykumar K. FTIR, XRD and DC conductivity of proton conducting gel polymer electrolytes based on polyacrylonitrile (PAN). International Journal of ChemTech Research;**6**(13):5214-5219

[19] Wei P, Hantao Z. Characterization PAN/ATO nano composites prepared by solution blending. Bulletin of Materials Science. 2008;**31**:807-811

[20] Pandey GP, Agrawal RC, Hashmi SA. Magnesium ionconducting gel polymer electrolytes dispersed with nanosized magnesium oxide. Journal of Power Sources. 2009;**190**:563-572

[21] Sreepathi RS, Rao KVS, Shareefuddin MUV, Subbarao and S. Chandra. Ionic conductivity and battery characteristics studies on PEO+AgNO3 polymer electrolyte. Solid State Ionics. 1994;**67**:331-334

[22] Jaipal RM, Sreekanth T, Chandrasekhar M, Subbrao UV. Ion transport and electro chemical cell characteristic studies of a new (PVP/Na NO3) polymer electrolyte syste. Journal of Materials Science. 2000;**35**:2841-2845

[23] Zurina O, Khairul BMI, Azhar A, Lisani O. A comparative study of lithium and sodium salts in PAN based ion conducting polymer electrolytes. Ionics. 2010;**16**:431-435

[24] Zang S, Lee JY, Hong L. Visualization of particle distribution in composite polymer electrolyte system. Journal of Power Sources. 2004;**126**:125-133

[25] Gurusiddappa J, Madhuri W, Padma Suvarna P, PriyaDasan K. Conductivity and dielectric behavior of polyethylene oxide-lithium per chlorate solid polymer electrolyte films. Indian Journal of Advances in Chemical Science. 2016;**4**: 14-19

[26] Narasimharao M, Vijayakumar K. Structural and conductivity studies of PAN-based $Al_2O_3$ nanocomposite gel polymer electrolytes. Iranian Journal of Materials Science and Engineering. 2018;**15**(4):11-18

[27] Mohamed NS, Zakaria MZ, Ali M, Arof AK. Electrochemical studies on polymer electrolytes based on poly(methyl methacrylate)-grafted natural rubber for lithium polymer battery. Journal of Power Sources. 1997;**66**:169-176

[28] Chandra A. Synthesis and ion transport characterization of hot-pressed $Ag^+$ ion conducting glasspolymer electrolytes. Journal de Physique. 2013;**87**:643-652

[29] Bhide A, Hariharan K. Ionic transport studies on (PEO) 6:NaPO3 polymer electrolyte plasticized with PEG400. European Polymer Journal. 2007;**43**: 4253-4270

[30] Bhargav PB, Mohan M, Sharma K, Rao N. Structural and electrical studies of sodium iodide doped poly(vinyl alcohol) polymer electrolyte films for their application in electrochemical cells. Ionics. 2007;**13**:173-182

[31] Krishna Jyothi N, Vijayakumar K. FTIR, XRD and DC conductivity studies of proton conducting gel polymer electrolytes based on polyacrylonitrile. International Journal of ChemTech Research. 2014;**6**:5214-5219

[32] Sreepathi RaoS. Rao K.V.S. Shareefuddin Md.U.V. Subbarao and S. Chandra., Ionic conductivity and battery characteristics studies on PEO+AgNO3 polymer electrolyte, Solid State Ionics, 67,331-334, 1994.

[33] Jaipal RM, Sreekanth T, Chandrasekhar M, Subbrao UV. Ion transport and electro chemical cell Characterstic studies of a new (PVP/Na NO3) polymer electrolyte system. Journal of Materials Science. 2000;**35**:2841-2845

[34] Lee JY, Kim DY, Kim CY. Synthesis of soluble polypyrrole of the doped state in organic solvents. Synthetic materials. 1995;7:103-106

[35] SreepathiRao S, Jaipalreddy M, Reddy KN, Subbarao UV. Solid State Ionics. 1994;**74**:225

[36] Naresh Kumar K, Sreekanth T, Jaipalreddy M, Subbarao UV. Journal of Power Sources. 2001;**101**:130

[37] Mohamed NS, Zakaria MZ, Ali AMM, Rof AKA. Journal of Power Sources. 1997;**66**:169

[38] Chetia JR, Maullick M, Dutta A, Dass NN. The effect of type of cation and salt concentration on ion–ion and ion–polymer interactions in PEG–MSCN (M = Li, Na, K) polymer electrolyte. Materials Science and Engineering B. 2004;**107**:134-139

[39] Sreepathi Rao S, Jaipalreddy M, Narasimha Reddy K, Subbarao UV. Conductivity study of polyethylene oxide (PEO) complexed with sodium bicarbonate. Solid State Ionics. 1994;**74**:225-232

[40] Naresh Kumar K, Sreekanth T, Jaipalreddy M, Subbarao UV. Study of transport and electrochemical cell characteristics of PVP: NaClO(3) polymer electrolyte system. Journal of Power Sources. 2001;**101**:130-139

[41] J Ramesh Babu, Ravindranath and Vijaya Kumar K, Structural and electrical studies of sodium citrate doped poly(vinyl alcohol) films for electrochemical cell applications"- Asian Journal of Chemistry, vol.29,No.5, Pg. 1049-1055, 2017

Chapter 8

# Phase Change Materials for Renewable Energy Storage Applications

*Banavath Srinivasaraonaik, Shishir Sinha and Lok Pratap Singh*

## Abstract

Solar energy is utilizing in diverse thermal storage applications around the world. To store renewable energy, superior thermal properties of advanced materials such as phase change materials are essentially required to enhance maximum utilization of solar energy and for improvement of energy and exergy efficiency of the solar absorbing system. This chapter deals with basics of phase change material which reflects, selection criteria, PCM works, distinguish thermal energy storage system, commercially available PCM, development of PCM thermal properties and durability of PCM. In addition to this chapter focused on PCM in solar water heating system for buildings particularly in India because 20–30% of electricity is used for hot water in urban households, residential and institutional buildings. Discussed Flat plate collectors (FTC) in detail which is suitable for warm water production in household temperature 55 to 70 °C owing to cost effective than the Evacuated Tube collectors (ETC), Concentrated collector (CC) and integration of different methods PCM in solar water heating system.

**Keywords:** Phase change materials, Advanced materials, Solar energy, Renewable energy

## 1. Introduction

Renewable energy is a free energy that can impact between energy supply and energy demand. One of the prominent renewable source is solar energy among the wind, rain, waves, tides and geothermal energy [1]. Most of countries receives 5 x $10^{15}$ kWh per annum i.e. incident mean solar energy in between 4 and 7 kWh per $m^2$ [2]. This can be accomplished in different solar energy fields such as solar water heating systems, desalination, solar-thermal collectors, building heating and day-lighting and Photovoltaic (PV) Cells etc. (**Figure 1**) [3]. Technologists and researchers are trying to utilize more renewable energy for distinguish devices/systems to decrease global energy crisis [4]. Thermal energy storage (TES) systems may assist the renewable energy exploitation for reduction of Green House gases (GHGs) and depletion of fossil fuels [5]. It plays vital role into energy conversation of free energy to reduce energy consumption [6]. TES can be stored in form of sensible heat, latent heat and thermochemical energy [7].

Sensible heat refers to the amount of energy is absorbed without phase change i.e. Solid–solid, liquid –liquid and gas to gas and latent heat refer to the amount of



IntechOpen

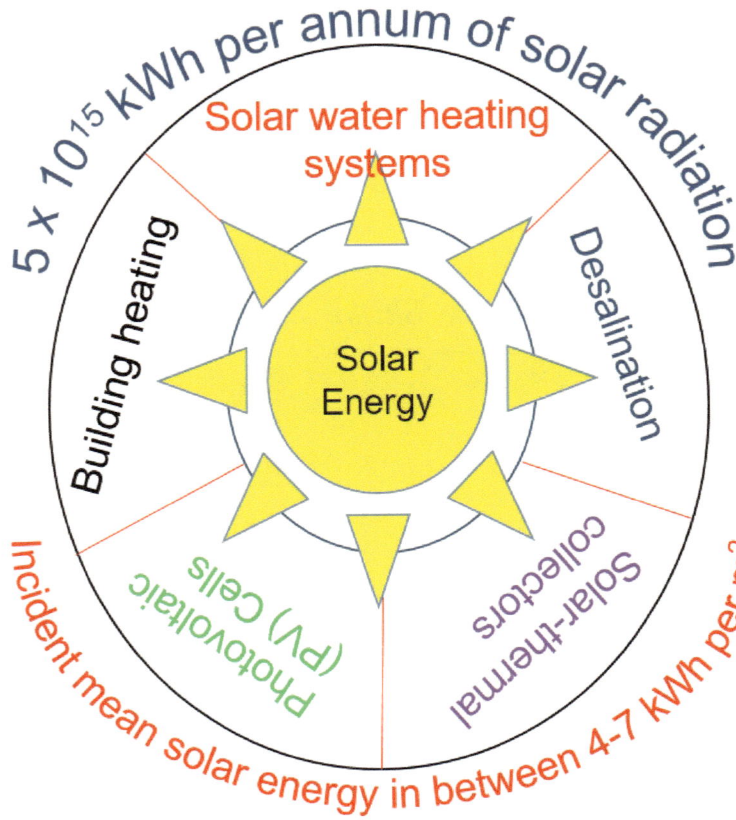


**Figure 1.**
*Application of solar energy in different fields.*

energy is absorbed with phase change i.e. solid to liquid, liquid to gas and solid to gas. Thermal chemical energy is energy stored during chemical reaction occurred not only at desired temperature range but also should be reversible reaction and Solid to liquid thermodynamically feasible to solar energy applications [8]. Solid to liquid materials are phase change materials (PCMs) and has the potential to store the energy at constant temperature owing to energy density per unit volume (**Figure 2**) [9].

Avargani et al. [10] installed two consecutive solar collectors with encapsulated paraffin phase change material. The single collector can produce hot water at the temperature of 60°C for 7 hour from midway to midnight. Fazilati and Alemrajabi [11] investigated the effect of PCM in solar water heater. The energy storage density was increased up to 39% and supply of hot water increased 25% as compared to the without PCM. Biwole et al. [12] PCM installed at the back of the solar collector. The solar collector was simulated using CFD model and compared with testing results. Added PCM to the back of solar collector can maintain the hot water temperature under 40°C for 80 min with the constant solar radiation of 1000 $W/m^2$. Hasan et al. [13–15] have incorporated different types fatty acids in domestic water heating system. The fatty acids such as myristic acid, palmitic acid and stearic acid, with phase transition between 50–70°C are the most promising PCMs for solar water heating. Manirathnam et al. [16] prepared nano-composite such paraffin wax as PCM with one per cent of Sci and CuO. Nano-composite, PCM and without PCM were studied in evacuated tube solar water heater for thermal energy storage. The energy efficiencies for distinguish cases were found to be 33.8%, 38.3%, and 41.7%, respectively corresponding to without PCM, PCM and nano-composite

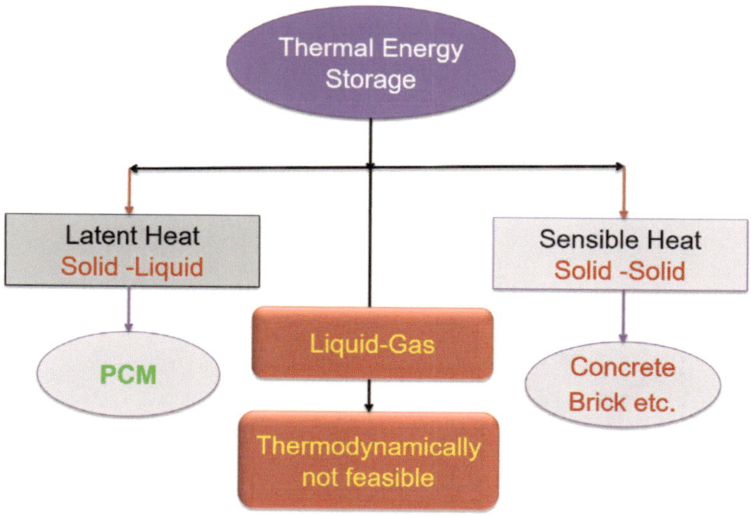


**Figure 2.**
*Types of thermal energy storage.*

respectively. Xie et al. [17] prepared cost effective and eco-friendly form shape stabilized stearic acid with coconut shell. The thermal properties of $SA/CSC_{15}$ composite were 76.69 J $g^{-1}$ and 52.52°C, respectively. The SA/CSC composite has potential for solar water heater energy storage.

This book chapter deals with basics of phase change materials and briefly discussed about selection criteria of PCMs. How these phase change materials are effective for solar water heater domestic uses as well as explained how low thermal conductivity of PCMs can be enhanced using supporting materials to increase efficiency of solar systems for thermal energy storage.

### 1.1 Working principle of phase change materials

When surroundings temperature above the PCM melting point, the PCM becomes phase change from solid to liquid and absorbs the heat from water storage tank during night, when surroundings temperature below the PCM melting point, the PCM desorbs heat to ambient/water storage tank, during the material changes phase from liquid to solid. The PCMs are being successfully used as energy storage devices such as heat pumps, solar engineering, space craft etc. [18].

When energy continues from the source, then PCM temperature rises from initial temperature ($T_1$) (K) to final temperature ($T_2$) (K) and during this period energy is riveted due to the sensible heat i.e. solid to solid [19]. The sensible heat can be calculated as per following Eq. (1)

$$Q_{\text{sensible heat}} = m.C_{ps}.(T_2 - T_1) \quad (1)$$

where Q is the amount of energy stored in the material (J), m is the mas of storage material (kg), $C_{ps}$ is the specific heat of the storage material of solid state (J/kg·K). From temperature ($T_2$) (K), the heat is continuously absorbed until the solid turns into the liquid due to the latent heat. The latent heat of system can be determined as per the following Eq. (2)

$$Q_{\text{latent heat}} = m.\Delta h \quad (2)$$

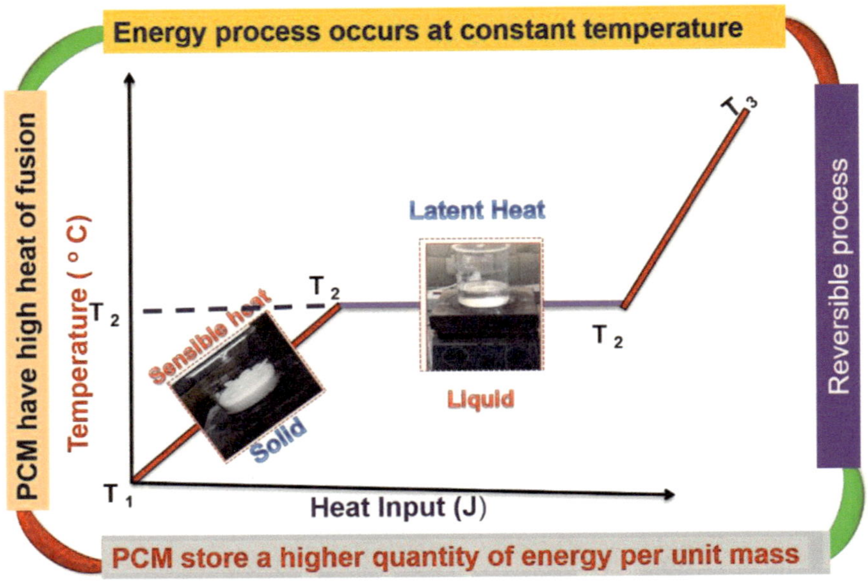


**Figure 3.**
*Working of phase change material.*

where Q is the amount of heat stored in the material (kJ), m is the mas of storage material (kg), and Δh is the phase change enthalpy (kJ/kg). Further, heat continues heat will be absorbed due to liquid to liquid. It means that, the amount of phase change materials need to be designed as per the application (**Figure 3**) [20].

$$\text{Total amount energy stored by PCM } (Q) = Q_{\text{sensible heat}} + Q_{\text{latent heat}} + m\, C_p l\, (T_1 - T_2) \quad (3)$$

$C_{pl}$ is the specific heat of the storage material of liquid state (J/kg·K).

## 2. Selection of phase change materials

Selection of PCMs for solar energy applications need to be considered the following properties.

### 2.1 Thermal point of view

PCMs should have high thermal conductivity during solid to liquid and liquid to solid for thermal cycling. PCMs should have high latent heat of fusion to store amount of energy required volume of the vessel prerequisite is less.

### 2.2 Physical point of view

PCMs should have high specific heat to absorb more heat during solid to solid i.e. sensible heat. PCMs should have high energy density per unit volume.

### 2.3 Kinetic point of view

PCMs should have high nucleation rate to avoid super cooling during liquid to solid. High crystal growth rate hassles for heat recovery.

### 2.4 Chemical point of view

PCMs should be reversible freeze/melt cycle. PCMs should be chemical stability i.e. functional groups contained in PCMs does not change after repeated thermal cycle. PCMs should be non-toxic, non-flammable and non-explosive materials for safety and Non-corrosiveness to the construction materials.

### 2.5 Economic point of view

PCMs should be easily available and low cost owing to minimize total cost of solar energy system.

## 3. Classification of phase change materials

Phase change materials are divided into three categories (i) Organic (ii) Inorganic (iii) Eutectic mixture.

### 3.1 Organic PCMs

The organic phase change compounds are chemically stable, no super cooling, non-corrosive and nontoxic. Organic PCMs are subdivided in two groups (i) Paraffins (ii) Non paraffins. Paraffins are chemically inert, have low thermal conductivity and large volume change. The non paraffin's such as fatty acids have high heat of fusion than paraffin and small volume change.

### 3.2 Inorganic PCMs

Inorganic PCMs have high heat of fusion, good thermal conductivity, are cheap and non-flammable. Most of them are corrosive to metals. Most inorganic PCMs are hydrated salt. Hydrated salts have a high energy density and high thermal conductivity. Disadvantage is that undergoes super cooling.

### 3.3 Eutectic mixture

Eutectic mixture is a mixing of more than one PCM material. Eutectic mixtures have sharp melting point and energy density is slightly higher than that of organic PCMs. Eutectics are divided in three groups (i) Organic – Organic (ii) Inorganic – Inorganic (iii) Organic – Inorganic [21]. The desired temperature range of eutectic mixture for solar energy applications can be designed according to Schroder's Eq. (3) [22].

$$\ln X_A = \frac{\Delta H_A}{R}\left(\frac{1}{T} - \frac{1}{T_f}\right) \tag{4}$$

Where $X_A$ and $\Delta H_A$ are the molar fraction and latent heat of fusion kJ/kg of compound A, respectively. T and $T_f$ are the melting temperature °C of the mixture and compound A. R is gas factor 0.8314 kJ/K. mol.

### 3.4 Bio-PCM

Bio-PCM is bio based materials which are derived from organic - based materials. It is less flammable than the commercial available PCMs. According to various

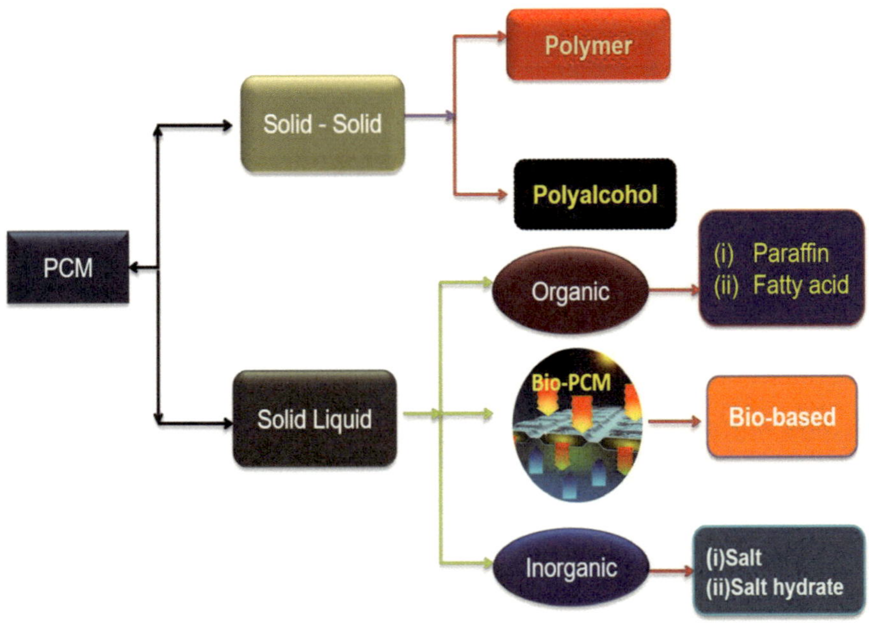


**Figure 4.**
*Classifications of PCMs [24].*

| Paraffin | Fatty Acid | Eutectic mixture |
|---|---|---|
| • Low thermal conductivity<br>• Low latent heat of fusion at desired temperature range. | • High thermal conductivity<br>• High Latent heat of fusion<br>• Small Volume Change | • Eutectics have sharp melting point similar to pure substance<br>• Volumetric storage density is slightly above organic compounds |
| **Disadvantages** | | |
| • Low thermal Conductivity<br>• Have large volume change | Lack of materials with phase transition around the thermal comfort | Only limited data is available on thermo-physical properties as the use of these materials. |

**Table 1.**
*Differentiate between raw PCMs.*

weather conditions, the bio-PCM can be prepared from - 22.7°C to 78.33°C. These materials are wraps in sheets as bubble. Bio-PCM has superior thermal properties such as specific heat and latent heat of fusion [23]. Classified PCMs are applying into different fields such as passive systems and active systems (**Figure 4**).

The active systems are waste heat recovery, solar water heater, desalination etc. The passive systems are directly added into the building components such as gypsum, mortar, concrete and brick. These two systems are employing in buildings to reduce energy burden in the buildings. Buildings are more responsible 40% of total energy consumption.

Comparison of different PCMS for renewable storage applications have provided in **Table 1** which helps for selection PCMs for thermal energy storage.

## 4. Availability of PCMs

In the market, it is available in encapsulated and un-encapsulated PCMs. The un-encapsulated PCMs such as Indiamart, Alibaba etc. are the manufacturing

**Figure 5.**
*Commercially available of encapsulated PCMs [30].*

companies. The encapsulated PCMs such as Microtek-BASF, Cristopia, Climator and Rubitherm are commercializing encapsulated PCMs (EPCMs) with the name of DS5001X, RT 5, and RT 25 etc. within temperature range of below ambient to above 100°C [25–29]. The encapsulated PCM is a tiny particle which contains core as PCMs and Shells are the polymers and inorganic substances (**Figure 5**).

## 5. Methods of incorporation of PCMs into renewable storage systems

PCMs can be incorporated into two ways one is macro-encapsulation and microencapsulation for thermal storage unit.

### 5.1 Macro-encapsulation

In macro-encapsulation methods, PCM has placed in size >1 mm. In this technique, a significant quantity of PCM can be packed in a closed container for subsequent used in thermal storage elements [30]. For better improvement of energy efficient, researchers are being positioned in various configurations such as Raw PCM in Metal ball, aluminum panels, Polypropylene flat panel, tube encapsulation. However, Metal ball and Aluminum panels have the superior thermal properties i.e. thermal conductivity than Polypropylene flat panel, tube encapsulation. It can be improved exergy and energy efficiency as well as duration of hot water outlet [31].

### 5.2 Microencapsulation

Encapsulation is a tiny particle has the particle size <1 mm where in PCM as a core material which is surrounded by inorganic shell such as Titanium, Silica etc. Polymers such as a Melamine – formaldehyde (MF), Urea Formaldehyde (UF), Poly-styrene (PS), Polyurethane (PU), Methyl methacrylate (MMA) etc. Microencapsulation of phase change materials can be prepared in two methods one is physical method and chemical method (**Figure 6**) [32].

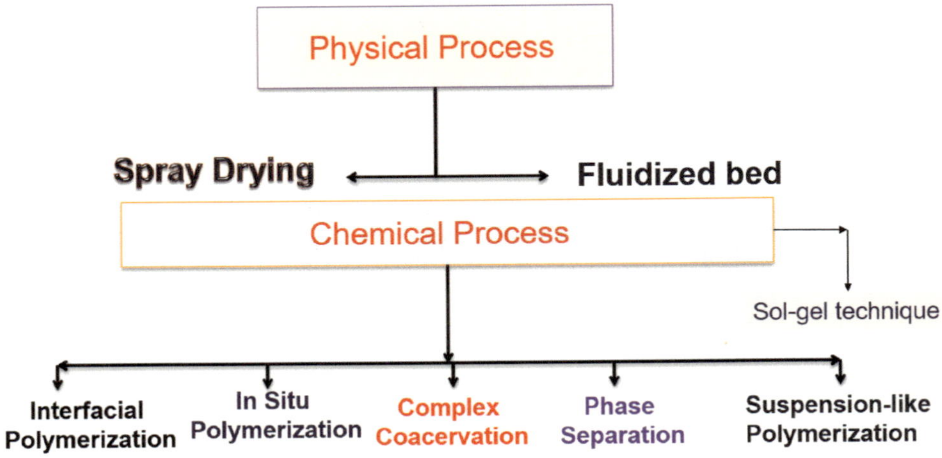


**Figure 6.**
*Methods for preparation of microencapsulation of PCM.*

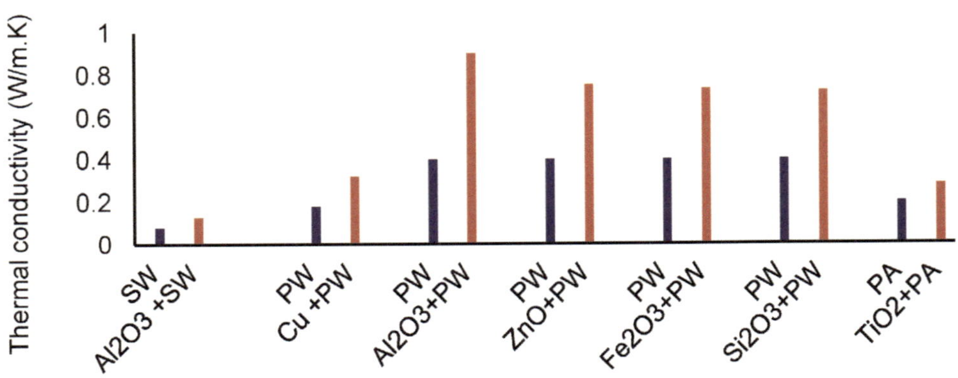


**Figure 7.**
*Improvement of effective thermal conductivity of PCMS for energy storage. *SW: Sugar WAX, PW: Paraffin wax, PA: Palmitic acid.*

This technique controls the volume change during solid to liquid, resists interaction with environment and enhances the heat transfer area [33]. The inorganic shells can improve the effective thermal conductivity of organic PCMs. Effective thermal conductivity is plays vital role in energy storage unit [34].

Addition of 2–4% of high thermal conductivity material to the PCM can be enhance its thermal properties [35] as shown in **Figure 7**. It will be helpful better performance of energy storage unit.

## 6. Durability of phase change materials

Accelerated thermal cycle test is essentially required for before applying in solar water heater and solar air conditioner. Thermal cycle test is referring to heating from ambient temperature to melting point of phase transition until completely becomes liquid and cools down to below melting point until becomes the solid. The total heating period and cooling period is called accelerated test. It works once in a day and reflects life of the phase change materials (**Figure 8**). Silakhori et al. [36] conducted accelerated test for paraffin wax and determined melting point and

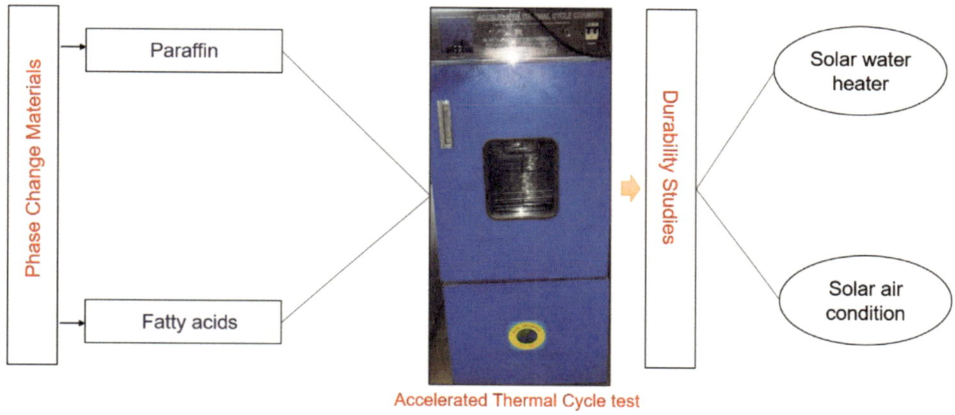


**Figure 8.**
*Performance of accelerated thermal cycle test.*

latent heat of fusion after 1000 cycles. 1.6–7% of melting point of paraffin wax was observed. Alkan et al. [37] conducted thermal cycle test of microencapsulated docosane for thermal stability with polymethyl methacrylate (PMMA). There is no significant changes occurred in key parameters 5000 cycles. Ahmet Sari et al. [38] performed the accelerated thermal cycling test for microencapsulated n-octacosane for 5000 cycles. There is no change observed in chemical structures of microcapsules. Sude Ma et al. [39] carried out conducted the thermal cycling test of paraffin wax with PMMA up to1000 cycles. No change in observed in thermal stability of microcapsules.

Yang et al. [40] performed accelerated test of different fatty acids such as lauric acid, myristic acid, palmitic acid and stearic acid for 10,000 thermal cycles. Thermal properties of fatty acids have not changed significantly after repeated cycles Sheili et al. [22] performed thermal cycle test of eutectic mixture of capric and lauric acid. no substantial changes in eutectic mixture after 360 cycles. Chinnasamy V and Appukuttan S [41] determined thermal properties of eutectic mixture of lauric acid /myristyl alcohol after 1000 cycles. It was determined that, there was no changes observed in thermal properties. Zuo et al. [42] found that eutectic mixture of lauric acid/1-tetradeconal were stable thermal properties were stable up to 90 thermal cycle tests. Zhang et al. [43] prepared ternary fatty acid mixture of PCMs with lauric acid, Mysteric acid, and palmitic acid. Melting point and heat of fusion were stable up to 50 cycles.

## 7. Phase change material for different solicitations for energy storage unit

Based on distinguish phase transition temperature range, these are incorporating in different solicitations are solar energy, building and vehicles for plummeting greenhouse gases (GHGs) and thermal management (**Figure 9**). The temperature ranges from −20°C to +5°C for domestic or commercial refrigeration. The second phase transition temperature range from +5°C to +40°C is applied for heating and cooling applications in buildings.

Now days, utilization of fossil fuels creates huge impact on environment and its leads to studies on commercial refrigeration, heating and cooling in building, solar heating and electronics, Textiles, building energy conversation etc. The PCMs

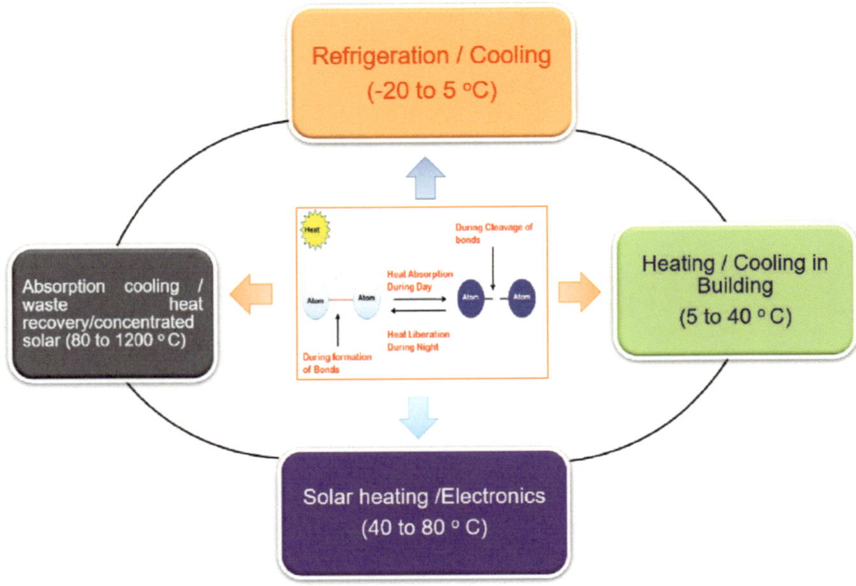


**Figure 9.**
*PCMs are in different applications.*

operating temperature range from +40°C to +80°C are used for solar based heating, hot water production and electronic applications and + 80°C to +1200°C range is applied for absorption cooling, waste heat recovery and concentrated solar [44–49].

## 8. Phase change material into the solar water heating systems

Solar radiation is occurred from the daylight and can be absorbed with solar collectors. These collectors are used for various applications; one of the solicitations is production of outlet hot water. The outlet of the hot water temperature is depending upon different types of collectors (**Figure 10**).

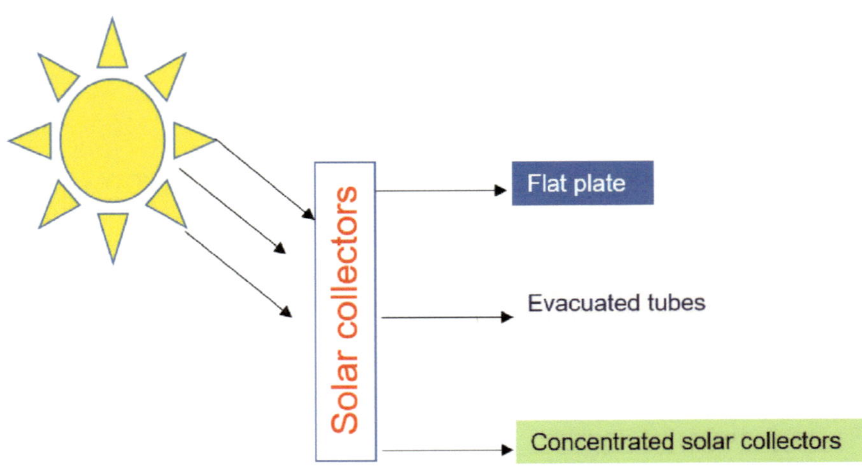


**Figure 10.**
*Different types of solar collectors.*

Generally, these solar collectors are mounted on walls for thermal management in the buildings. The thermal power output of the various solar collectors can be determined with product of conversion efficiency and intensity of solar irradiance [44]. The output of thermal power collector can be calculated using following Eq. (4)

$$Q_{KN}=\left(n_0-\frac{\alpha_1(\theta_K-\theta_u)+(\theta_K-\theta_u)^2}{E}\right).E.A_K=m.C_p.(\theta_{KO}-\theta_{KI}) \quad (5)$$

$Q_{KN}$ Output of thermal power of collector (W), E solar irradiance intensity (W / $m^2$), $A_K$ Collector area ($m^2$). Where: $n_0$: Zero-loss collector efficiency, $\alpha_1$: Basic heat loss coefficient (W/$m^2$ K), $\theta_{K}$: Mean collector temperature (K), $\theta$ u: Ambient air temperature (K).$\theta_{KO}$: Collector outlet temperature (K), $\theta_{KI}$: Collector inlet temperature (K), m: HTF mass flow rate (kg/s), $C_p$: heat capacity of HTF (J/kg K).

Among all the solar collector, the flat plate solar collector is discussed in detail owing to Manufacture process is easy, cost effective, maintenance is low and easy installation. This type of Flat plate solar water heater is suitable for urban households (**Table 2**).

## 8.1 Flat plate

Flat plate is one type of heat exchanger for solar collector that converts radiant energy from sunlight into heat energy. This plate is generally used for low and moderate temperature applications i.e. <80°C. This type of collector contains one is case, second is absorbers like copper or aluminum positioned in the heat exchanger owing to good conductors for heat, the heat transfer fluid and insulation materials. To improve thermal efficiency, need to be minimizing the thermal losses and integrate the superior thermal properties of PCM. The thermal storage materials can be integrated either in the collector or separate thermal storage tank. Flat plate collectors are used in hot water production and space heating, and air conditioning system [50, 51]. For solar water heating, the flat-plate collectors are installed at the optimum angle is Latitude +10°. Water is transport fluid in solar water heating and it has good thermodynamic properties such as high heat capacity, high energy density and incompressible.

Disadvantage of water as transport fluid is damage the collector when it is freeze in winter. The damage can be managed by positioning collector at low solar inputs and need to be add antifreeze mixtures to improve above mentioned problems. The usual antifreeze substances are ethylene glycol or propylene glycol. These chemicals are variegated with water and proper discarding due to toxicity. The durable of antifreeze chemicals is about 5 years [52].

In Flat plate solar water heater, PCM can be equipped in two ways (i) Flat plate integrated solar collector (ii) Flat plate non-integrated solar collector.

| Solar collector type | Solar collector efficiency | Heat loss coefficient ($W/m^2.K$) |
|---|---|---|
| Unglazed absorber | 0.91 | 12.0 |
| Flat-plate, single glazing, non-selective absorber | 0.86 | 6.1 |
| Flat-plate, Single glazing, selective absorber | 0.81 | 3.8 |
| Flat-plate, double glazing, selective absorber | 0.73 | 1.7 |
| Vacuum | 0.80 | 1.1 |

**Table 2.**
*Performance data for distinguish solar collectors [45].*

### *8.1.1 Flat plate integrated solar collector*

In Flat plate integrated solar collector, PCM can be positioned like aluminum honeycomb structure and PCM modules for frost protection under the absorber plate (**Figure 11**). PCM integrated solar collectors are increased thermal stability and extended for hot water outlet. However, advanced insulation materials are to be attached to minimize the heat losses otherwise may reduce efficiency of the system [53].

### *8.1.2 Flat plate non: integrated solar collector*

In Flat plate non – integrated solar collectors connected to PCM storage unit. The PCM storage unit is placed on upper of an inclined collector, near the solar collector or under the solar collector. To avoid the leakage, the PCMs are encapsulated in rectangular, cylindrical, and spherical container (**Figure 12; Table 3**).

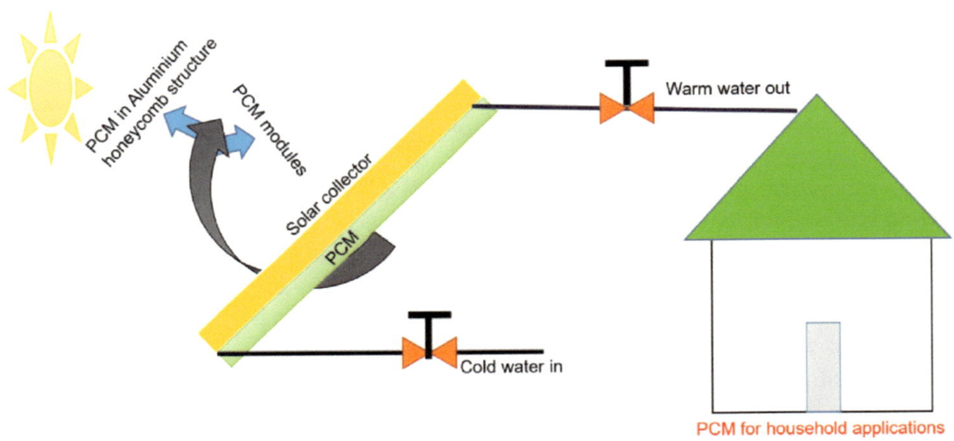


**Figure 11.**
*PCM in flat integrated solar collector.*

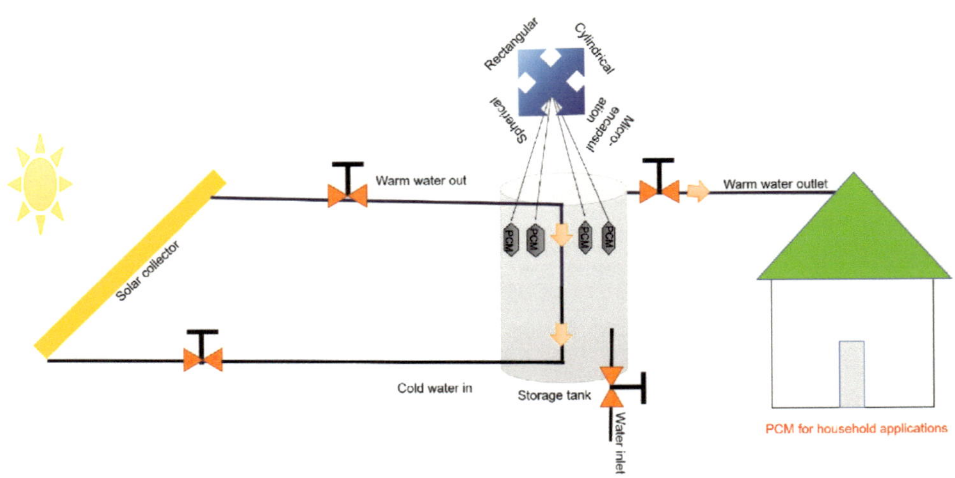


**Figure 12.**
*PCM in in flat non-integrated solar collector.*

| Types | Advantages | Disadvantages |
|---|---|---|
| Rectangular | • Manufacturing process is easy<br>• Small occupied space | • Heat loss rate is high<br>• Thermal stress concentration<br>• Leakage phenomenon may exist |
| Cylindrical | • Fluid flow can improved<br>• Rate of the PCM is high | • Manufacturing process is not easy |
| Spherical | • High heat transfer efficiency<br>• Low heat loss rate | • Positioned in storage is complex<br>• Complicated filling process of the PCM is difficult |
| Micro-encapsulation | • Encapsulation efficiency is high<br>• Particle size is low<br>• High heat transfer area is high | • Manufacturing process is tough<br>• Manufacturing costs is easy |

*M.P: melting point, H.F.: Heat of fusion, SA: Stearic acid, MA: Myristic acid.*

**Table 3.**
*Merits and demerits of regular PCM container for different medium [54].*

Incorporating solar energy storage system into building may diminish cost of renewable energy storage system and also progress efficiency of the collection. In solar water heating process, the storage unit is filled with PCM for captivating the heat during day from hot water. At night, the absorbed energy supplies to the warm water tank and hot water can be collected for a long time [55, 56]. Kulakarni and Deshmukh [57] studied efficiency of water heating system using paraffin whose melting point was 62°C. Efficiency of solar water heater increased from 31.25% to 44.63%. The storage capacity was enhanced from 3260.4 kJ to 4656.5 kJ. Bhargava [58] utilized three different thermal properties of PCMs such as $Na_2SO_4$. $10H_2O$ (32 °C and 251 kJ/kg), $Na_2HPO_4.12H_20$ (36.1°C and 279 kJ/kg) and P116 (46.7°C and 209 kJ/kg) Wax were incorporated into storage unit. Determined the efficiency of the system and duration of the outlet water temperature. As thermal conductivity of the materials are increased, Increased duration of outlet hot water temperature during the evening hours. Fazilati and Alemrajabi [59] used Paraffin as storage medium. The melting point and latent heat of fusion were 55°C and 187 kJ/kg. 39%, 16% and 25% improved the energy and exergy efficiency and duration of warm water was improved. Prakash et al. [60] laminated a PCM layer (46.7°C and 209 kJ/kg) at the bottom of the water tank. They concluded that, it was not effective during phase change from liquid to solid due to low heat transfer area. Kaygusuz [61] had studied performance of solar water with $CaCl_2.6H_2O$ (28°C and 45 kcal/kg) as phase change material an experimental and theoretically. Hasan et al. [13–15] incorporated some fatty acids as PCMs such as myristic acid (MA), palmitic acid (PA) and stearic acid (SA) for domestic water heating. They recommended that these fatty acids with melting temperature between 50–70°C were the most auspicious PCMs for water heating. Most of the researchers were studied with different phase transition temperature in solar water heater system. However, as per the Cabinet of Ministers of Latvia, the allowable domestic hot water (DHW) range must be from 55 to 70°C [62]. Literature review given in **Table 4** on phase transition temperature rage in 55 to 70 °C for DHW.

Other than above PCM, some of commercial available within 55–70°C of thermal storage materials are listed out in Zalba et al. [78]. These materials may have applied in Flat Plate Solar water heater for better thermal efficiency, thermal management and longer duration for warm water.

| PCM | M.P (°C) | H.F (kJ/kg) | Reference |
|---|---|---|---|
| Two kinds of PCM | 70 | 210 | [63] |
| Paraffin | 60 | 213 | [64] |
| Paraffin and SA | 61 & 57 | 213 & 198 | [65] |
| Salt hydrate | 60 | | [66] |
| RT 60 | 60 | 144 | [67] |
| Nano Cu-PCM (0.5 to 2%) | 57.81–59.57 | 157.3 to 172.2 | [68] |
| RT 65, SA, Pent glycerin | 55,66,80 | 159,207,152 | [69] |
| RT 65 graphite composite | 65 | — | [70] |
| Paraffin | 70–80 | 224 | [71] |
| PCM1 | 60–62 | 209 | [72] |
| Paraffin | 60–70 | 224 | [73] |
| — | 57.34 | 178.76 | [74] |
| MA, Paraffin, Tristearin | 58, 59,56 | 199,189,191 | [75] |
| Sodium acetate tri-hydrate with graphite | 60 | 180–200 | [76] |
| SA-MA (80–20%) | 61–65 | 190.87 | [77] |

**Table 4.**
*Literature review on PCM flat plate solar collector for water heater.*

## 9. Conclusion

Phase change materials have high energy density and potential to apply in Flat plate solar collector for production of hot water in urban households. Other than the researchers attempted, there are so many PCMs available commercially in the market for improvement of efficiency of Solar water system. Thermal cycle test is essential for determining durability of paraffin and fatty acid before applying into system. Paraffin and fatty acid have durability to 14 years and 27 years respectively. Higher thermal conductivity of PCMs increased longer duration of hot water however low thermal conductivity materials with high latent heat of fusion of PCMs can be enhanced with addition of high thermal conductivity fillers. Encapsulated PCMs is a tiny particle can easily have applied in storage tanks. PCMs are successfully incorporated in integrated and non-integrated flat plate solar collector. However, non-integrated flat plat solar collector has the higher thermal efficiency than the integrated solar collector because of difference in heat transfer area. There are number of PCMs, commercial methods and designs are available at national and international level. Among them cost effective parameters are selected for effective PCM solar water heating system.

## Abbreviations

| | |
|---|---|
| Q | amount of energy stored in the material (J) |
| m | mass of storage material (kg) |
| $C_{ps}$ | specific heat of the storage material of solid state (J/kg·K) |
| $T_1$ | Initial temperature (K) |

| | |
|---|---|
| $T_2$ | Final temperature (K) |
| $\Delta h$ | phase change enthalpy (kJ/kg) |
| PCM | Total amount energy stored (Q) |
| $C_{pl}$ | specific heat of the storage material of liquid state (J/kg·K). |
| $X_A$ and $\Delta H_A$ | molar fraction and latent heat of fusion kJ/kg of compound A |
| T and $T_f$ | melting temperature °C of the mixture and compound A |
| R | gas factor 0.8314 kJ/ K. mol |
| $Q_{KN}$ | Output thermal power of collector (W) |
| E | Solar irradiance intensity ($W/m^2$) |
| $A_K$ | Collector area ($m^2$) |
| $n_0$ | Zero-loss collector efficiency |
| $\alpha_1$ | Basic heat loss coefficient ($W/m^2$ K) |
| $\theta_K$ | Mean collector temperature (K) |
| $\theta u$ | Ambient air temperature (K) |
| $\theta_{KO}$ | Collector outlet temperature (K) |
| $\theta_{KI}$ | Collector inlet temperature (K) |
| m | HTF mass flow rate (kg/s) |
| $C_p$ | Heat capacity of HTF (J/kg K) |

## Author details

Banavath Srinivasaraonaik[1*], Shishir Sinha[2] and Lok Pratap Singh[1]

1 IBM Group, CSIR – Central Building Research Institute, Roorkee, Uttarakhand, India

2 Department of Chemical Engineering, Indian Institute of Technology, Roorkee, India

*Address all correspondence to: srinivas@cbri.res.in

## References

[1] Sharma A, Chen CR: Solar Water Heating System with Phase Change Materials. International Review of Chemical Engineering. 2009;1: 4.

[2] Jamil B, Siddiqui AT, Akhtar N: Estimation of solar radiation and optimum tilt angles for south-facing surfaces in Humid Subtropical Climatic Region of India. Engineering Science and Technology, an International Journal. 2016;19 : 1826-1835.

[3] Yakup M, Malik AQ. Optimum tilt angle and orientation for solar collector in Brunei Darussalam. Renewable Energy. 2001;24 : 223–234.

[4] Dharma S, Masjuki HH, Ong CH, Sebayang HA, Silitonga SA, Kusumo F, Mahlia IMT. Optimization of biodiesel production process for mixed Jatropha curcas-Ceiba pentandra biodiesel using response surface methodology. Energy Convers. Manag. 2016; 115: 178–190.

[5] Sharma A, Tyagi V, Chen C, Buddhi D: Review on thermal energy storage with phase change materials and applications. Renew. Sustain. Energy Rev. 2009;13: 318–345.

[6] Murray RE, Groulx D. Experimental study of the phase change and energy characteristics inside a cylindrical latent heat energy storage system: Part 1 consecutive charging and discharging. Renewable energy. 2014; 62: 571-581.

[7] Cot-Gores J, Castell A, and Cabeza LF. Thermochemical energy storage and conversion: A-state-of-the-art review of the experimental research under practical conditions. Renewable and Sustainable Energy Reviews. 2012; 16 : 5207–5224.

[8] Cabeza LF, Martorell I, Miró L, Fernandez AI, Barreneche C: Introduction to thermal energy storage (TES) systems. Advances in Thermal Energy Storage Systems. 2021;1-33.

[9] Silakhori M, Jafarian M, Arjomandi M, Nathan GJ: Comparing the thermodynamic potential of alternative liquid metal oxides for the storage of solar thermal energy. Sol. Energy. 2017; 157: 251–258.

[10] Avagani VM, Norton B, Rahimi A, Karimi H. Integrating paraffin phase change material in the storage tank of a solar water heater to maintain a consistent hot water output temperature. Sustainable Energy Technologies and Assessments. 2021 ; 47: 101350.

[11] Fazilati AM, Alemrajabi AA. Phase change material for enhancing solar water heater, an experimental approach. Energy Conversion and Management. 2013;71: 138–145.

[12] Biwole PH, Eclache P, Kuznik F. Phase-change materials to improve solar panel's performance. Energy and Buildings. 2013 ; 62: 59–67.

[13] Hasan A. Thermal energy storage system with stearic acid as phase change material, Energy Conversion and Management. 1995; 35: 843–856.

[14] Hasan A. Phase change material energy storage system employing palmitic acid, Solar Energy 1994; 25; 143–154.

[15] Hasan A, Sayigh A. Some fatty acids as phase change thermal energy storage materials. Renewable Energy. 1994; 4: 69–76.

[16] Manirathnam AS, Manikandan MKD, Prakash RH, Kumar BK, Amarnath MD. Experimental analysis on solar water heater integrated with nano composite phase change material (Sci and CuO). Materialstoday. 2021;37 : 232-240.

[17] Xie, Baoshan, Li, Chaunchang, Zhang, Bo, Yang, Lixin, Xiao, Guiyu, Chen, Jian. Evaluation of stearic acid/ coconut shell charcoal composite phase change thermal energy storage materials for tankless solar water heater. Energy and Built Environment. 2020; 1:187-198.

[18] Fang G, Li H, Liu X and Wu S. Preparation and characterization of nano-encapsulated n-tetradecane as phase change material for thermal energy storage. Chemical Engineering 2009; 53: 217-221.

[19] Mehling H, Cabeza LF. Heat and Cold Storage with PCM: An Up to Date Introduction into Basics and Applications. Heidelberg, Berlin: Springer. 2008.

[20] Zalba, B., Marín, J.M., Cabeza, L.F. and Mehling H. Review on thermal energy storage with phase change: materials, heat transfer analysis and applications. Applied Thermal Engineering. 2003; 23: 251–283.

[21] Batens R, .Jelle BP, Gustavsen A Phase change materials for building applications: A state- of-the –art review. Energy and Buildings.2010;42: 1361-1368.

[22] Shilei L, Neng Z, Guohui F. Eutectic mixtures of capric acid and lauric acid applied in building wallboards for heat energy storage. Energy and Building. 2005; 38: 708-711.

[23] Muruganantham K, Phelan P, Horwath P, Ludlam D, McDonald T. Experimental Investigation Of A Bio-Based Phase-Change Material To Improve Building Energy Performance. Proceedings of ASME. 2010:4 th International Conference on Energy Sustainability ES2010.

[24] Shchukina EM, Graham M, Zheng Z, Shchukin DG. Nano encapsulation of phase change materials for advanced thermal energy storage systems. Chem. Soc. Rev. 2018 ;47: 4156 —4175.

[25] www.cristopia.com.

[26] www.teapcm.com.

[27] www.climator.com.

[28] www.rubitherm.de.

[29] www.basf.com

[30] Waqas A, UdDin Z. Phase change material (PCM) storage for free cooling of buildings — A review" Renewable and Sustainable. Energy Reviews. 2013; 18: 607–625.

[31] Schossig P, Henning HM, Gschwander S, Haussmann T. Micro-encapsulated phase change materials integrated into construction materials. Solar Energy Materials and Solar Cells. 2005; 89; 297-306.

[32] Borreguero AM , Carmona M, Sanchez ML , Valverde JL, Rodriguez FJ. Improvement of thermal behaviour of gypsum block by the incorporation of microcapsules containing PCMS obtained by suspension polymerization with an optional core/coating ratio. Applied Thermal Engineering. 2010; 30: 1164-1169.

[33] Fang Y, Kuang S, Gao X, Zhang Z. Preparation and characterization of novel nanoencapsulated phase change materials. Ener. Convers. Manage. 2008; 49: 3704 – 3707.

[34] Srinivasaraonaik B, Singh LP, Sinha S, Tyagi I, Mittal G. Studies on thermal properties of microencapsulated eutectic phase change material incorporated different mortar mixes. International Journal of Energy Research. 2020; https://doi.org/10.1002/er.5943

[35] Tangsiriratana E, Skolpap W, Patterson RJ, Sripapha K. Thermal

properties and behavior of microencapsulated sugarcane wax phase change material. 2019;5: 02184.

[36] Silakhori M, Naghavi MS, Simon H, Metselaar C, Meurah T, Mahlia I, Fauzi H, Mehrali M. Accelerated Thermal Cycling Test of Microencapsulated Paraffin Wax/ Polyaniline Made by Simple Preparation Method for Solar Thermal Energy Storage Materials. 2013; 6: 1608-1620.

[37] Alkan C, Kaya K, SarI A. Preparation, thermal properties and thermal reliability of form-stable paraffin/polypropylene composite for thermal energy storage. J. Polym. Environ. 2009; 17: 254–258.

[38] Sari A, Alkan C, Karaipekli A, Uzun O. Microencapsulated n-octacosane as phase change material for thermal energy storage. Solar Energy. 2009; 83: 1757–1763.

[39] Ma SD, Song GL, Miao ZC, Wang DW. Preparation and characterization of paraffin/PMMA core/ shell structured microcapsules. Mater. Adv. Res. 2011; 239–242, :524–527.

[40] Yang Li, Cao Xi, Zhang N, Xiang Bo, Zhang Z, Qian B. Thermal reliability of typical fatty acids as phase change materials based on 10, 000 accelerated thermal cycles Sust. Cities and Soc. 2019;46:101380.

[41] Chinnasamy V, Appukuttan S. Preparation and thermal properties of lauric acid/myristyl alcohol as a novel binary eutectic phase change material for indoor thermal comfort, Energy Storage 2019;1:e80.

[42] Zuo J, Li W, Weng L. Thermal properties of lauric acid/1-tetradecanol binary system for energy storage Appl. Therm. Eng. 2011; 31:1352-1355.

[43] Zhang N, Yuan Y, Wang X , Cao X, Yang X, Hu S. Preparation and characterization of lauric–myristic–palmitic acid ternary eutectic mixtures/ expanded graphite composite phase change material for thermal energy storage Chem. Engg. J. 2013; 231: 214–219.

[44] Douvi E, Pagkalos C, Dogkas G, Koukou K. M, Stathopoulos N. V, Caouris Y, Michail Gr. V., Phase Change Materials in Solar Domestic Hot Water Systems: A review, Int. of therm. Fluids. 2021;2: 100075.

[45] CEN, Solar energy - Solar thermal collectors - Test methods (ISO 9806: 2013). Brussels: CEN. 2013.

[46] Sreekumar C, Veerakumar A. Phase change material based cold thermal energy storage: Materials, techniques and applications – A review. International Journal of Refrigeration. 2016; 67: 271-289.

[47] Gholamibozanjani G, Farid M. Application of an active PCM storage system into a building for heating/cooling load reduction. Energy. 2020; 210:118572.

[48] Sardari PT, Roohollah BM. Energy recovery from domestic radiators using a compact composite metal Foam/PCM latent heat storage. Journal of Cleaner Production. 2020;257:120504.

[49] Qin D, Yu, Z Yang T, Li S, Zhang G. Thermal performance evaluation of a new structure hot water tank integrated with phase change materials. Energy Procedia. 2019; 158: 5034-5040.

[50] www.alternative-energy-tutorials.com

[51] Badiei Z, Eslami M, Jafapur K. Performance Improvements in Solar Flat Plate Collectors by Integrating with Phase Change Materials and Fins: A CFD Modeling. Energy. 2020; 192: 1-15.

[52] Kalogirou SA. Solar energy engineering: Processes and Systems. Elsevier. 2009

[53] Abuska M, Sevik S, A. Kayapunar A. Experimental Analysis of Solar Air Collector with PCM-Honeycomb Combination under the Natural Convection. Solar Energy Materials and Solar Cells. 2019;195: 299-308.

[54] Ling X, Tian L, Yang L, Yifei L, Qianru L. Review on application of phase change material in water tanks. Advances in Mechanical Engineering. 2017;9:1-13.

[55] Jouhara H, Khordehgah N, Almahmoud S, Delpech B, Chauhan A, Tassou S. Waste heat recovery technologies and applications. Thermal Science and Engineering Progress. 2018; 6: 268-289.

[56] Tailor M, Patel Y, Sindha U. Solar Water Heater Using Phase Change Material. IJARIIE.2017;3: 2395-4396.

[57] Kulkarni MV, Deshmukh DS. Improving Efficiency of Solar Water Heater Using Phase Change Materials. IJSSBT. 2014; 3: 2277-7261.

[58] Bhargava AK. Solar water heater based on phase changing material. Applied Energy. 1983;14:197–209.

[59] Fazilati MA , Alemrajabi AA. Phase change material for enhancing solar water heater, an experimental approach, Energy conversion and management. 2013; 71: 138-145.

[60] J. Prakash, H. P. Garg, G. Datta, A solar water heater with a built – in latent heat storage, Energy Conversion and Management 25 (1985) 51-56.

[61] Kaygusuz K. Experimental and theoretical investigation of latent heat storage for water based solar heating systems. Energy Conversion and Management. 1995; 36: 315-323.

[62] Dzikevics M, Veidenbergs I, Valancius K. Sensitivity Analysis of Packed Bed Phase Change Material Thermal Storage for Domestic Solar Thermal System. Environmental and Climate Technologies. 2020; 24: 378-391.

[63] Zhou F, Ji J, Yuan W, Zhao X, Huang S. Study on the PCM Flat-Plate Solar Collector System with Antifreeze Characteristics. International Journal of Heat and Mass Transfer 2019; 129: 357-366.

[64] Nallusamy N, Sampath S, Velraj R. Experimental investigation on a combined sensible and latent heat storage system integrated with constant/varying (solar) heat sources. Renewable Energy. 2007; 32: 1206–1227.

[65] Reddy R, Nallusamy N, Reddy K. Experimental studies on phase change material-based thermal energy storage system for solar water heating applications. Renewable Energy Applications. 2012; 2:1-6.

[66] Dzikevics M, Veidenbergs I, Valancius K. Sensitivity Analysis of Packed Bed Phase Change Material Thermal Storage for Domestic Solar Thermal System. Environmental and Climate Technologies. 2020; 24: 378-391.

[67] Elbahjaoui R, El Qarnia H. Thermal Performance of a Solar Latent Heat Storage Unit Using Rectangular Slabs of Phase Change Material for Domestic Water heating Purposes. Energy and Buildings. 2019; 182: 111-130.

[68] Saw C, Al-Kayiem H. Evaluation of copper nanoparticles – Paraffin wax compositions for solar thermal energy storage. Solar Energy. 2016;132: 267–278.

[69] Haillot D, Nepveu F, Goetz V, Py X, Benabdelkarim M. High performance storage composite for the enhancement of solar domestic hot water systems. Part 1: Storage material investigation. Solar Energy. 2011; 85: 1021-1027.

[70] Haillot D, Nepveu F, Goetz V, Py X, Benabdelkarim M. High performance storage composite for the enhancement of solar domestic hot water systems. Part 2: Numerical system analysis. Solar Energy. 2012; 86: 64-77.

[71] Chen Z, Gu M, Peng D. Heat transfer performance analysis of a solar flat-plate collector with an integrated metal foam porous structure filled with paraffin. Applied Thermal Engineering. 2010; 30:1967–1973.

[72] Yang L, Zhang X, Xu G. Thermal performance of a solar storage packed bed using spherical capsules filled with PCM having different melting points. Energy and Buildings. 2014; 68: 639-646.

[73] Chen Z, Gu M, Peng D, Peng C, Wu Z. A Numerical Study on Heat Transfer of High Efficient Solar Flat-Plate Collectors with Energy Storage. International Journal of Green Energy. 2010; 7: 326-336.

[74] Shirinbakhsh M, Mirkhani N, Sajadi B. Optimization of the PCM-Integrated Solar Domestic Hot Water System under Different Thermal Stratification Conditions. Energy Equipment and Systems. 2016; 4: 271-279.

[75] Shirinbakhsh M, Mirkhani N, Sajadi B. A Comprehensive Study on the Effect of Hot Water Demand and PCM Integration on the Performance of SDHW System. Solar Energy. 2018; 159: 405-414.

[76] Cabeza L, Ibanez M, Sole C, Roca J, Nogués M. Experimentation with a water tank including a PCM module. Solar Energy Materials and Solar Cells. 2006; 90: 1273-1282.

[77] Mazman M, Cabeza L, Mehling H, Nogues M, Evliya H and Paksoy H. Utilization of phase change materials in solar domestic hot water systems. Renewable Energy. 2009; 34: 1639-1643.

[78] Zalba B, Marin MJ, Cabeza FL, Mehling H. Review on thermal energy storage with phase change: materilas, heat transfer analysis and applications. Applied thermal engineering. 2003; 23: 251-283.